ISO 14001 CERTIFICATION

**Prentice Hall International Series
in The ISO 14000 Environmental Management
System Standards**

The purpose of the series is to present essential information and tools necessary to do practical environmental management in compliance with the ISO 14000 Standards. The series will consist of a set of books, some with accompanying software. Written for professionals working in the field and for students, the books in the series will be complete in their coverage of the ISO 14000 Standards, covering the topics of *Environmental Management Systems—14001, Environmental Auditing—ISO 14010s, Environmental Labeling—ISO 14020s, Environmental Performance Evaluation—ISO 14030s,* and *Life-Cycle Analysis—ISO 14040s.* Although the topics are presented in the important context of the ISO 14000 standard, the environmental subjects and tools covered in the books are practical and useful environmental management principles, independent of ISO certification.

Current Publications:

ISO 14001 Certification—Environmental Management Systems,
publication date July 1995

ISO 14010s—Environmental Auditing,
publication date December 1995

Other publications to be announced in 1996 and 1997.

ISO 14001 CERTIFICATION

Environmental
Management Systems

*A Practical Guide for Preparing Effective
Envionmental Management Systems*

W. Lee Kuhre

Seagate Technology, Inc.,
Senior Director Environmental Health and Safety

University of San Francisco
Graduate Program Senior Lecturer

For book and bookstore information

http://www.prenhall.com

Prentice Hall PTR, Upper Saddle River, NJ 07458

Library of Congress Cataloging-in-Publication Data

Kuhre, W. Lee, 1947–

   ISO 14001 Certification : environmental management systems

   Lee Kuhre.

       p. cm.

   Includes bibliographical references and index.

   ISBN 0-13-199407-7 (case bound)

   1. Industrial management—Environmental aspects.   2. Environmental protection—Standards.   I. Title.

HD30.255.K84   1995   95-17082

658.4'08—dc20   CIP

Editorial/Production Supervision and Interior Design: Lisa Iarkowski
Chapter Opener and Icon Art: Gail Cocker-Bogusz
Acquisitions Editor: Bernard Goodwin
Manufacturing Manager: Alexis R. Heydt
Cover Design: Design Source

 © 1995 Prentice Hall PTR

Prentice-Hall, Inc.

A Paramount Communications Company

Upper Saddle River, NJ 07458

The publisher offers discounts on this book when ordered in bulk quantities.
 For more information, contact:

   Corporate Sales Department
   PTR Prentice Hall
   One Lake Street
   Upper Saddle River, NJ 07458
   Phone: 800-382-3419
   FAX: 201-236-7141
   e-mail: corpsales@prenhall.com

Printed in the United States of America

10 9 8 7 6 5 4 3 2 1

ISBN 0-13-199407-7

Prentice-Hall International (UK) Limited, London
Prentice-Hall of Australia Pty. Limited, Sydney
Prentice-Hall of Canada, Inc., Toronto
Prentice-Hall Hispanoamericana S.A., Mexico
Prentice-Hall of India Private Limited, New Delhi
Prentice-Hall of Japan, Inc., Tokyo
Simon & Schuster Asia Pte. Ltd., Singapore
Editora Prentice-Hall do Brasil, Ltd., Rio de Janeiro

Dedicated to
Margot
&
Marcail

Table of Contents

PART TWO: *Initial Environmental Elements/Components Needed* **45**

PART THREE: *Environmental Management Programs and Operational Control Procedures* 91

List of Tables

Preface

There have been many significant events in the field of environmental management, however, most pale in comparison to the addition of environmental management into the International Organization for Standardization (ISO) certification system. This monumental event is expected to occur in 1996, with many organizations starting to prepare now. A great amount of effort is presently being spent by many committees around the world to establish this new environmental management set of standards that will be entitled the ISO 14000 series.

New regulations that require environmental management systems are implemented almost daily in most countries. It has always been a challenge for the environmental manager to decide which systems to design and how to implement them in a cost-effective way. When you add the comprehensive environmental management systems that will now be required for ISO 14001 certification, the task for the environmental manager could easily become overwhelming. This book was prepared to offer practical guidance concerning selection, design and implementation of the environmental management systems necessary to achieve ISO 14001 certification which in turn will help meet the regulations.

Once the standard is in place, it is expected that customers will require their product suppliers to be ISO 14001 certified. In addition, some service suppliers who have an impact on the environment will also probably be expected to obtain an ISO 14001 certification. Certification will imply to the customer that when the product or service was prepared, the environment was not significantly damaged in the process. Most customers already demand this type of certification now in terms of quality under the ISO 9000 certification system.

In recent years Great Britain has developed many comprehensive standards for universally acceptable and certifiable quality standards. These standards are described in British Standard (BS) 5750 which later resulted in ISO 9000 standards. Many organizations in Europe, the United States and various other countries are now ISO 9000 certified or in the process of becoming certified.

To complement this, BS 7750 was written to establish environmental management on the same basis. Significant portions of BS 7750 are presently being incorporated into the new environmental management standard ISO 14000 series by Technical Committee 207 of the ISO. Many countries are represented on TC 207 including the United States.

The impact will be significant on the environment and organizational dynamics. The environment will benefit because most organizations will have to prepare and implement significantly more comprehensive environmental management systems than they presently have in order to obtain certification. This means considerable additional time and money to upgrade environmental management systems to a level that will allow ISO 14001 certification.

Organizations must now include care of the environment in their everyday operations. Customers of certified organizations will be assured that the products or services they purchase have been produced in accordance with universally accepted standards of environmental management. Organization claims, which today can be misleading or erroneous, will, under the standards, be backed up by comprehensive and detailed environmental management systems which must withstand the scrutiny of intense audits.

Organizational costs to achieve certification will vary widely. Those who have basically sound and complete policies and programs for quality and environmental control may need only minor change or adjustment for certification. Others who are in the start-up phase of operations or who have not yet established effective environmental policies or programs, will find the process expensive.

This book discusses the scope of the proposed environmental management standards and offers boilerplate-type procedures to obtain certification under them. The first five chapters present background. Chapters 6–10 present the initial elements and components needed. Chapters 11–21 contain information about the actual procedures and programs that an organization will have to prepare and implement in order to obtain ISO 14001 certification.

The icon appearing in the margin throughout the book indicates when there is an easy to use boilerplate-type procedure available in *Appendix F* and on the floppy disk accompanying this book.

The availability and cross-referencing of these boilerplate procedures and word processor templates can be summarized as follows:

Word Processor Template Index Name	Appendix F—Title of Procedure	Applicable Chapter
Initial Word Processing Guidance Files		
1-tips.doc or .wp	Suggestions on How to use the Floppy Disk	
2-index.doc or .wp	Cross-Reference/Index	
3-format.doc or .wp	Blank Template	
Boiler Plate Environmental Procedure Templates		
approval.doc or .wp	Process, Equipment and Chemical Approvals	12
auditing.doc or .wp	Audit, Review and Verification Procedure	14
awarenes.doc or .wp	Internal Communication & Employee Awareness	7
chemcont.doc or .wp	Handling of Empty Chemical Containers	12
chemtrck.doc or .wp	Tracking of Chemicals	12
closure.doc or .wp	Site Closure	21
communic.doc or .wp	Procedure for External Communication	17

Filenames that end in .doc are available on the floppy disk in Microsoft Word for Windows 6.0, and files that end in .wp are available in Word Perfect 5.1.

The author wishes to thank some very important people who assisted with the preparation of this book. Thanks to Marjorie Kuhre for editorial review; to Andrew Allan, Jr., for technical input, to Margot and Marcail Kuhre for being supportive daughters, to Dr. Joseph Petulla (University of San Francisco) and Eve Levin (Region 9 EPA) for technical review, to Michele Corash, Esq. for legal input; to Robb Kundtz (Seagate Technology) for offering his encouragement; to Lisa Iarkowski for production supervision and to Mike Hays of Prentice Hall for suggesting the book be written; and to Bernard Goodwin for agreeing to publish the book.

A few citations, especially in Chapter 4, are referenced as U.S. Sub Tag 1. This refers to unpublished, non-copyrighted ISO 14001 U.S. Sub-tag 1 paper work. The author has gone to great lengths not to repeat or duplicate the information in important ISO 14000 committee documents which will later be published by the committees. This was done in order to respect the hard work of those committee members who have donated so much of their personal and organization's time to make the ISO 14000 standards the best they can be. The author thanks and acknowledges these many dedicated men and women.

The author has tried to be as complete and accurate as possible. Considering the fact that environmental management is an extremely broad and complex topic, it is possible that errors and omissions have occurred. The reader should also refer to final ISO 14001 standards and the applicable international, federal, state, regional and local regulations. The author, Prentice Hall, Seagate Technology and the University of San Francisco cannot be held responsible for any errors or omissions.

About the Author

W. Lee Kuhre, Senior Corporate Director of Environmental, Health, Safety and Security for Seagate Technology, oversees environmental management at sites in 21 countries with over 50,000 employees. He is also a Graduate Senior Lecturer at the University of San Francisco. He is a Certified Hazardous Materials Manager, a Registered Environmental Assessor, a Registered Environmental Professional, and is AHERA Certified. He is author of *Practical Management of Chemicals and Hazardous Wastes* (Prentice Hall).

Part I

Background

The first five chapters in this book contain information necessary to understand generally what ISO14000 and environmental management are all about in this new era of ecological consciousness. These chapters set the stage for organizations that are considering improving their environmental management systems and/or obtaining ISO 14001 certification. Once these basics have been covered in Part I, the actual elements, programs and operational controls will be given in the chapters in Parts II and III, which will help the reader obtain an ISO 14001 certification and a comprehensive environmental management system.

Chapter 1
Introduction

Significance

The addition of environmental management into International Organization for Standardization (ISO) certification is monumental in significance. The impact will be felt in terms of the environment, operation of most organizations, customers, agencies and most components of society. There will be a positive impact on the environment since many organizations will have to start doing additional environmental protection to obtain certification. Most organizations that want to obtain or maintain ISO certification will have to spend time to implement additional comprehensive environmental management programs, even if they already have many in place. This is especially significant to customers as they will be better able to evaluate whether the product or service they are buying was prepared in an environmentally conscious way. Agencies should also benefit in that more of their requirements should be met indirectly because of ISO 14001 certification. All in all, this is an extremely significant event in the history of environmental management and the worldwide impact it will have.

To set standards for environmental management that can be applied worldwide for most organizations seemed like an impossible task. The environment is just too different 50 feet away, and even more so across the world, and to apply a standard that works for widely different industries, such as a small clothing manufacturer and an oil refinery, used to be unthinkable. But the standard has been created and for organizations not to prepare for it would be shortsighted.

In order to get a feeling for the significance of ISO 14000, a comparison to the current ISO 9000 quality certification is helpful. Anyone who has obtained ISO 9000 quality certification can attest to the tremendous amount of time and effort that was needed to obtain and maintain certification. Some organizations have set up an entire staff with huge operating budgets for ISO 9000 alone, even if they already had good quality systems in place. Many feel that the addition of environmental management to ISO will be just as involved as it has been for quality. The customer now knows that ISO 9000 certification represents massive effort and implies good quality, but was the environment destroyed in the process? Many customers want the same level of assurance in terms of the environment as they have with quality.

Overview

The subjects in this book are organized into three parts. Part I, Chapters 1 through 5, presents background information necessary to understand the ISO 14000 Environmental Management System (EMS) process. The initial "hands-on" elements and procedures required for ISO 14001 certification or just plain good environmental management are presented in Part II, Chapters 6 through 10. Chapters 11 through 21 are organized into Part III and provide techniques and procedures that should be prepared after the initial systems are set up.

The addition of environmental management into ISO certification has been, and will continue to be, a very complicated endeavor and is therefore hard to summarize briefly. In general, however, it is the identification of environmental management systems, policies and programs that organizations should establish and maintain in order to protect the environment. Since operating conditions vary so greatly from one industry to another and in different parts of the world, agreement on one set of standards that everyone can understand and accept was difficult.

Figure 1–1 illustrates the environmental management process and components that must be addressed in order to obtain certification. One of the most important aspects of the figure is the feedback nature, with one component continually feeding into the next, thus allowing continu-

OVERVIEW OF MAJOR COMPONENTS OF ENVIRONMENTAL MANAGEMENT NEEDED FOR ISO 14001 CERTIFICATION

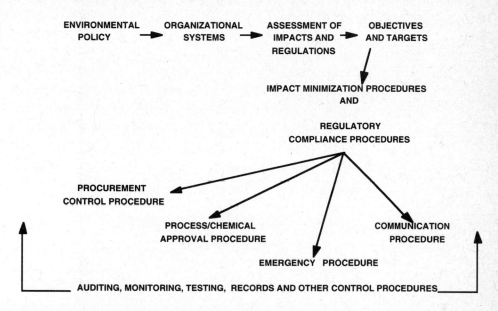

Figure 1–1

ous improvement. Training, communication and documentation are key components since they impact all the other system elements. Life-cycle analysis, environmental labeling, environmental performance and environmental aspects of product standards are related subjects but will probably not be required for certification. There are many other aspects of environmental management which are not shown in this figure but which are presented in the rest of this book. It is only possible to show a few of the highlights or core elements, each of which can be subdivided. The entire process must be able to continuously adjust and improve since the environment and most organizations are not static.

The most commonly asked ISO 14000 questions deal with how closely an organization must follow the specifications and what the document(s) must look like. Based upon the way that ISO 9000 quality certification occurred in most cases, it is expected that if the basic procedures are in place and the intent is met in a consistent way, environmental certification will be obtained. In other words, if environmental impacts are minimized by consistent procedures that are routinely followed, certification will probably be obtained. This may occur without creating massive new documents.

Another question that many organizations are asking concerns where the emphasis should be directed: toward writing or implementing. Without question it is better to overimplement, not overdocument. If too many procedures and documents exist, there will be less progress in terms of concrete actions taken to help the environment. If a manageable number of environmental procedures exist which are understood and implemented by the employees, the true intent is being met.

A clear distinction needs to be made between ISO 14000 and agency regulations. ISO 14000 deals with procedures and systems, not agency-mandated discharge standards or limitations. For example, agency regulations already address the number of parts per million (ppm) that can be discharged to the air, water or land on a case-by-case basis. ISO 14000 will not repeat or conflict with the agencies. ISO 14000 will instead require that certain procedures or components of environmental management systems, such as targets and objectives, be in place. The actual targets and their levels are determined by the organizations and agencies. In other words organizations will still be required to satisfy agency laws and regulations. ISO 14001 certification is not required by agencies; however, some customers may require certification.

Many environmental management systems are known by different names. The important thing is that a consistent system is there, regardless of what it is called. Certification can probably be obtained even if the program titles are not identical to those mentioned in the standards. For example, many refer to these subjects as environmental control, pollution prevention, waste minimization, environmental engineering and numerous other names. The systems might be part of other documents or programs that relate directly to the business of the organization. This is not only acceptable from an ISO 14000 standpoint but preferable as well. If environmental management is deeply integrated into the operation, it will be able to make much more meaningful and long-term positive changes.

History

Two historical evolutions are especially important in this subject: environmental management and standardization. Environmental management has existed in some form for thousands of years but really started in earnest in the 1960s. Significant contamination discovered at that time prompted the enactment of many laws and regulations in the 1970s and 1980s. Waste minimization became a popular component of environmental management in the late 1980s and early 1990s.

Standardization on a worldwide basis was accelerated with quality and occurred generally independent of environmental management. The work of Deming and other quality experts started getting considerable attention in the 1980s. Attempts to "standardize" quality requirements were made by many organizations; however, it wasn't until the 1990s that considerable agreement was reached. One standard that received significant attention was British Standard 5750, which in large part led to the current ISO 9000 quality standard in which most organizations worldwide are investing incredible resources to achieve.

The environmental management and the standardization movements merged in the early 1990s. This has occurred through the hard work of many individuals and organizations such as the International Organization for Standardization (ISO), the British Standards Institute (BSI), the American National Standards Institute (ANSI), and numerous other organizations in many countries.

There are many proposed standards involved in this merger of environmental management and standardization. Figure 1–2 illustrates some of these standards, such as BS7750, that are being considered. Technical Committee (TC) 207 to ISO also has a draft standard which has a high probability of being adopted. Many different organizations and individu-

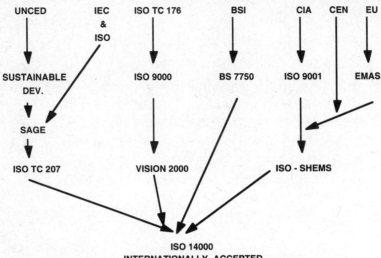

EVOLUTION OF ISO 14000

ABBREVIATIONS:
UNCED = 1992 United Nations Conference on the Environment and Development
SAGE = 1991 Strategic Advisory Group for the Environment
ISO = International Organization for Standardization
TC = Technical Committee
BSI = British Standards Institute
BS = British Standards
CIA = Chemical Industries Association
SHEMS = Safety, Health, and Environmental Management Systems
EU = European Union
EMAS = Eco-Management and Audit Scheme
CEN = Comite European de Normalisation
IEC = International Electrotechnical Commission

Figure 1–2

als around the world have worked on the proposed standards in an attempt to make them reasonable and acceptable to all the countries that have endorsed ISO 14000.

The major difference between the BS 7750 and TC 207 drafts is that BS 7750 generally requires more. For example, under BS 7750 impacts must be disclosed to the public. There are also tighter requirements for continuous improvement of environmental performance, use of best available technology (BAT) and use of performance standards for auditing (Cutter Information Corp., 1994). The standards being developed by TC 207 are not as rigorous as BS 7750; however the annex to the TC 207 standards includes most of the items mentioned above as suggested programs. At the time of publication of this book it appears that the TC 207 standards will be the ones adapted for worldwide application.

The remaining chapters of this book present background details, procedures and suggestions for meeting these standards. In fact, many of the generic procedures could be used as a starting point. This would prevent re-creating the wheel from scratch. Site-specific information should be incorporated into these suggested general procedures.

Chapter 2
Benefits
of ISO Certification

Introduction

There are numerous benefits of ISO 14001 certification. Most of these could also be considered benefits of environmental management. The most important benefit is protection of the environment. Following the suggestions presented will also help demonstrate compliance with regulations and the establishment of effective management systems. These are practical management systems designed to minimize environmental impacts in a cost-effective manner.

Protection of the Environment

A core purpose of ISO 14001 certification is to allow plants and animals to continue to exist in the best conditions possible. ISO certification in environmental management may only be one small step; however, the process will grow and improve as experience is accumulated. Creation, documentation and maintenance of the systems needed for certification can only help the environment.

Possibly the greatest positive impact to the environment will be in the reduction of hazardous waste. ISO certification requires programs which will reduce hazardous materials (chemicals) and hazardous waste. These types of programs result in less hazardous waste needing land disposal, which in turn results in less soil and ground-water pollution. Not only waste management, but also waste minimization are key parts of ISO certification. Waste minimization is the way of the future in terms of environmental protection and applies to hazardous and nonhazardous waste.

Nonhazardous waste minimization will also have a dramatically positive impact on the environment and should be a key component of the plans prepared to achieve ISO certification. This would apply to reduction, reuse or recycling, all of which maximize natural resources. Paper, cardboard and aluminum are usually reduced, reused or recycled more than most other wastes because there is normally a market for the recycled materials. Waste minimization of plastic and the other types of solid waste is also a benefit to the environment and should be a part of the ISO certification program. Unfortunately, most organizations have to pay to have these types of wastes recycled.

Another environmental benefit would be conservation of other natural resources. For example, a good environmental management program will help reduce the need for electricity, gas, space and water and therefore conserve these valuable commodities. In some parts of the world these resources are in short supply and conservation programs are already in place. Even if they are, there is always more that can be done to conserve or reduce resource use. This not only helps the environment but usually results in a cost savings for the organization as well.

ISO 14001 certification could also be a common platform to aid in the correction of certain worldwide environmental problems. Some environmental issues, such as depletion of the ozone layer, cross all geographical boundaries. ISO 14000 could help in the management of these types of international or global issues.

Equal Competitive Basis

At the present time there is no uniformity in the amount of time and money that organizations currently spend to protect the environment. This is true even within the same industry in the same geographical area. Certification will help put organizations on a more equal basis in terms of the time and money they spend to help protect the environment. In today's competitive market is it fair for one company to spend $.50 on the dollar when a competitor is spending $.05 on the dollar and polluting the environment in the process? It will never be possible to completely equalize environmental control costs; however, ISO 14001 certification will at least reduce the larger discrepancies between competitors and therefore give the environmentally conscious organization more of a chance.

Demonstrated Compliance with Regulations

Most organizations that have been in existence for a few years are probably in partial compliance with most of the applicable regulations. If they were not, chances are they would be out of business by now because of lawsuits, negative publicity and multiple other side effects. Fines assessed can even bankrupt an organization. By having the ISO certification for environmental management there is a good chance that the written documentation necessary to demonstrate compliance with the regulations will be present.

Currently most agencies are basically watching the ISO 14000 movement and not becoming deeply involved. Regardless of this, it is probable that when an organization shows that they are ISO certified in terms of environmental management, agencies may not investigate as deeply. Since the ISO auditor has spent considerable time auditing an organization, it will not be necessary for every applicable agency to devote as much energy to assure that an organization is in compliance with the regulations.

The Environmental Protection Agency (EPA) does presently have representatives on some of the ISO 14000 committees. For example, the ISO 14000 SC1 to the TC207 committee for the United States has representatives from several EPA regions including the Washington D.C. area. The input from this office has been valuable, especially since most of the representatives for this particular subcommittee are from industry.

Overall, agency/industry relations should improve after ISO 14001 certification. The agency will know the certified organization cares about the environment and has systems in place, even before visiting the operation. This positive predisposition is extremely valuable and should help foster a better working relationship.

Establishment of Effective Management Systems

Effective management is good business, as are the planning, documentation and execution of environmental management systems. With as many variables as there are in keeping an organization viable, the quality, and now the environmental management standards, will force management to be more effective.

The standards that must be met to achieve ISO certification in terms of environmental management contain many good management techniques. Environmental personnel management, accounting, vendor control, document control and many of the other required systems are common management systems with an environmental twist. For example, environmental personnel training is especially important since the field is changing at such a rapid pace. Therefore, a training development program should be specified in general and specifically for each employee in this field. A training program is not unique to good environmental management since it should be done in most other fields as well.

Due to the variety of environments, regulations, problems and management techniques, it is easy for a professional in the field to become overwhelmed. This book presents some of the more "tried and true" environmental management systems that are realistic and cost-effective. It is important for the environmental management professional to use their time efficiently since there are usually far too many problems, opportunities, crises and regulations to address. It is necessary to prioritize the list of issues to tackle since it would be physically impossible to resolve them all. The prioritization should be adjusted frequently.

There are an almost infinite number of tools which the environmental manager can utilize. A few stand out, however, as especially practical. At the top of the list would be regulation databases, services or summaries. Hazardous waste tracking in computer software systems helps manage data concerning the volume, type, vendor, cost and other details. Most of the rest of the chapters in this book present the practical environmental management tools that will help achieve compliance with the regulations and ISO auditors.

Reduced Cost

After some initial costs have been incurred to design and implement missing programs and to obtain certification, there should be long-term cost savings, especially in the area of environmental control and cleanup. It has been estimated that over $100 billion is spent annually in the United States to comply with federal regulations pertaining to environmental control and pollution cleanup. This figure was estimated by the U.S. Environmental Protection Agency (NCMS, 1994). Certification will not eliminate all cleanup costs, however, it should minimize the number and size of future cleanups.

If initial and ongoing certification costs are properly managed, there should be a long-term reduction in environmental costs and increase in competitive edge. Since most organizations will be incurring the initial costs, there should not be a serious effect on an organization's competitive status. The costs will be partially offset by increased customer satisfaction, trust in the organization and higher morale. Those organizations that become certified early will have a powerful public-image advantage which can impact cost in a positive way. Innovation and ingenuity of employees in this work can provide a company with an edge in lowering costs to meet the standards. Hence, these qualities should be sought in the search for employees and encouraged in existing employees.

Once certification has been obtained, less time and money will have to be spent responding to customer questions. Presently most customers have a different and lengthy list of environmental questions. To show the organization's ISO 14001 certification will probably be enough for most customers.

The primary basis of the cost savings will be due to less chemicals and wastes handled or cleaned up. Fewer chemicals, fewer poor grade chemicals, fewer chemical spills and less hazardous waste which must be tracked and disposed of properly would be involved. Ground-water cleanups will be minimized.

Another reason for cost savings is due to the philosophy of doing it right the first time. This concept applies to ISO 9000 and numerous other applications as well. In terms of environmental management it is especially important, however, since doing it wrong one time can result in monumental cost and impact (Wortham, 1993).

Reduced Injuries

As these systems reduce the amount of chemicals and hazardous wastes on-site, the number of employees injured by these substances will also decrease. Obviously, to prevent suffering and possible death is a great benefit. The costs associated with injuries will also be reduced. Many improvements are extremely hard to quantify, especially in terms of injury costs. When management systems are implemented, they will not only protect lives, but they will also reduce the costs associated with loss of productivity and morale, equipment replacement costs and loss in time to correct and prevent recurrence.

Since environmental management and employee health and safety are so closely related, when one field benefits from a major change, it is highly probable that the other will as well. Systems that protect or minimize impacts on the environment will in most cases also minimize impact on the employees. This equates to reduced employee injuries and illness. After all, the employee is really part of the overall environment. A reduction in injury and illness will occur if the organization includes health and safety in its ISO 14001 certification efforts in addition to environmental management. It is the author's recommendation that health and safety be included in ISO 14000 efforts for the established organization (operated for at least five years) and not for the start-up organization. For the new organization, addressing more than just environmental management (and quality) might be biting off more than they can chew.

Improved Community Relations

The majority of the public really does care about the environment these days. It was found in a 1994 Gallup poll that citizens in 24 industrialized and developing nations consider environmental protection to be more important than economic growth (NCMS, 1994). If an organization improves their environmental management program, it will surely improve their community relations as well whether they obtain an ISO 14001 certification or not.

There are numerous environmental actions an organization can take to maintain or improve community relations and credibility. No major spills would be one of the best things an organization should strive to achieve. Secondly, waste minimization can also help improve community relations and is a key component of the ISO certification. Everything presented in this book would help in terms of community relations as it all deals with environmental management.

Most of the procedures that ISO 14000 requires are proactive environmental actions. Any proactive action that an organization does is good for the environment and could be communicated to the public since it is a positive story. If the public is aware of these things, their confidence in the organization will be increased.

Improved Customer Trust and Satisfaction

Closely related to improving community relations is the subject of customer trust and satisfaction. Once an organization has the ISO 14001 certification, the customer can't help but feel more secure that the environment is being protected. It will assure the customer that the supplier really does care about the environment. This sense of environmental security will be based upon a certification which is a little more tangible than the lip service given in many cases in the past. Previously some companies made statements about protecting the environment that had no basis. Numerous package labels, TV and magazine advertising and other unsupported environmental statements resulted in low customer trust. With ISO 14000, an organization can now assure their customers and the general public that they really are protecting the environment and they have adequate documentation to back up the statement. All the customer should need to hear is that their supplier is ISO 14001 certified.

In some parts of the world it has gone even past customer trust and satisfaction. For example, some U.K. customers (such as the U.K. Ministry of Defense) are reportedly now requiring an early form of BS 7750/ ISO 14001 certification of their suppliers (Burhenn, 1994). Whether it is required or not, an ISO 14001 certification will provide a valuable competitive edge. Organizations that obtain the certification early should be able to increase their market share for their product since most customers are environmentally conscious these days.

Improved Upper Management Attention

The whole process of obtaining ISO 14001 certification will give upper management in most organizations an increased and more positive appreciation for environmental management (Wortham, 1993). This type of attention has been needed in many organizations for quite some time and is even missing in many cases. In the past the environmental department has been regarded by many individuals in the organization as a necessary evil or a cost burden. With ISO 14000, the environmental department will be regarded as a positive and important component of the team. This will happen since the customer is now actively interested and involved in the environment.

Chapter 3
Application and Scope

Introduction

This chapter expands the introduction presented in Chapter 1 and provides more of the detail concerning the who, what, when and where of ISO 14000 and environmental management in general. After reading this chapter, the reader will clearly know whether the book applies to them or not and whether to continue reading. It is probable that ISO 14000 and/or environmental management apply to 99 percent of all readers. This is due to the fact that most organizations, even very small ones, have a measurable impact on the environment.

Who

The inclusion of environmental management standards in the ISO system should be of interest to anyone who is responsible for or cares about environmental management or is concerned with certifications. In other words, anyone who sells a product or certain services may have to deal with this issue at some point in time, if not now. This would apply to almost any type of organization, whether an industry, a utility, consulting firm, public agency or a service or product supplier.

In terms of size, customers seem to be concentrating initially on the larger organization in terms of ISO certification; however, all organizations with 10 or more individuals will probably have to become certified at some point in time. Size is really not important since some small organizations can have significant impacts on the environment, sometimes greater than large organizations.

Organizations needing certification will be in the design, start-up, operational or shutdown modes. It really doesn't matter what stage the organization is in as far as certification. If they are presently producing a product or service, certification is probably already required or soon will be. If the organization is in the planning stage, they might as well start introducing the systems required for certification now or they may never obtain customers.

Those who are currently certified for quality (ISO 9000) have taken the first step toward environmental certification in that quality standards and environmental standards share numerous common elements. In addition ISO 9000 certified organizations already possess an internal structure that will allow for faster certification. Appendix B to the EMS specification (SC1) lists the common sections of both ISO 9000 and 14000. By using this matrix it is possible to reuse many procedures, such as for training, that have already been prepared.

ISO 14000 could apply to parts of the organization, to the entire organization and to the contractors as well. The annex does point out, however, the choice is up to the organization. For example, only certain operations or all components within the organization may choose to obtain certification. Obviously the customer may have something to say about this. It is expected that many customers will want their entire supplier's organization certified.

The organizations that visually impact the environment in an obvious way will be under the most pressure to obtain ISO certification in terms of environmental management. These will probably include mining, petroleum refining, chemical manufacturing, power generation and heavy construction. Electronics manufacturing and telecommunications may be next, followed by retailers, suppliers and practically all other organizations.

What

Overall Environmental Management

ISO 14000 is a voluntary environmental management standard; however, some customers may require it to do business. Many aspects of environmental management must be included to obtain certification. In an over-simplistic way of thinking, it is the design and implementation of an environmental management framework to minimize the impact of the operation on the environment. In many cases the existing environmental documents can be used in whole or in part, along with the required new systems. Examples of environmental management systems that will be needed include policies, organizational systems, management, planning, operational procedures, effect and regulatory identification procedures, objectives, targets, vendor controls, auditing, record keeping and many others that will be discussed later.

The depth or complexity of the environmental management systems needed will depend on many things. Location, type and complexity of operation, level and number of environmental impacts and operating conditions are a few variables that determine the depth needed. If an organization is having many impacts on the environment, they will need a much more in-depth system than an organization with little impact. In addition to impact, the number of employees will also suggest a reasonable level of effort or depth to the ISO auditor. In terms of regulatory compliance, however, the number of employees does not correlate to depth of the environmental management systems needed.

It is important for the reader to understand that this book will concentrate on general environmental management systems and tools, which comprise the scope of ISO 14000. It will not deal with emission level criteria, performance standards, regulatory limits or other numerical parameters that are more appropriately covered in legislative type documents.

Total Quality Environmental Management and Continuous Environmental Improvement

Two terms that are gaining popularity help describe what will be required to achieve certification. These terms are total quality environmental management (TQEM) and continuous environmental improvement (CEI). Both have slightly different meanings, however, each helps demonstrate the spirit of ISO 14000. Everything presented in this book, beginning with Chapter 6, are most of the elements needed to achieve and live out TQEM, CEI and ISO 14001 certification.

TQEM brings the concepts of quality and environmental management together. If things are done right the first time and followed through the entire evolution of the product with quality of the environment in mind, everyone benefits. This idea applies as well to environmental management as it does to quality of products. If environmental considerations are incorporated at the beginning of an organization and product and carried through to the end, the environment and the organization will win.

Continuous environmental improvement is similar in concept to TQEM. It presents the concept that systems can always be improved, even after considerable resources have been expended and impacts are under control. There will always be a more cost-effective way to reduce environmental impacts even further, as long as there are creative individuals in the organization who are allowed to express their ideas.

Proactive Versus Reactive

One very important aspect of ISO 14000 is that it forces organizations to become more proactive. Far too often environmental individuals spend much of their time dealing with environmental crises, such as spills or discharges which are out of compliance. The process of becoming certified will require that more energy be redirected toward proactive program design and implementation. In most organizations there will have to be some individuals who are reassigned to prepare all the documents required for ISO 14001 certification and to help with implementation. This process is in fact a proactive course of action since it prepares the organization to handle environmental issues in the future as well as the present.

Products and Services

ISO 14000 applies especially to product manufacturers; however certain service providers may be required to be certified by their customers as well. Manufacturers of products that utilize chemicals will probably be under the greatest pressure to become certified. Almost all other manufacturers will then need to consider certification since they impact the environment at least in terms of energy, water and other natural resources.

Service providers whose impact on the environment has been recognized by others outside of the organization should consider certification as well. In addition to helping the environment, it will also help the service provider's reputation and marketing efforts. Hazardous waste transporters and treatment, storage and disposal facilities (TSDFs) are examples of service providers that will need ISO 14001 certification before most other service providers. Environmental consulting firms, utilities and many other service providers should also consider ISO 14001 certification.

Level of Detail

In most subject areas it is quality not quantity that is important. ISO 14000 is no exception. It is better to have fewer procedures that are consistently implemented or practiced than to have a truckload of procedures that are followed only in part or not at all. An organization would not want to go to the other extreme, however. Environmental management is a complicated subject requiring adequate volume to address all of the important systems and procedures.

Appearance of the Documents

What should the documents look like or how should all the paperwork be organized? Will auditors "judge a book by its cover"? The answers to these questions really depend upon each particular auditor. It is recommended, however, that there should not be numerous binders with ISO on the title. Organizations should utilize existing environmental management procedural documents they have already set up to achieve compliance with agency regulations and then supplement them only where they do not meet ISO 14000 specifications.

In certain parts of the world, especially in the United Kingdom, there may be some auditors who lean toward the requirements of BS 7750 in terms of documents, even if BS 7750 was not adapted completely. This standard requires that at a minimum a Register of Regulations, an Environmental Effects Register and an Environmental Manual are available (Rothery, 1993). In the United States it is felt that most auditors will not need to see numerous binders created solely for ISO 14000. As stated earlier, what really is important is the presence of the basic environmental management systems and procedures in some form and their implementation.

Referencing within Documents

Since there will not have to be binders labeled "ISO 14000," it is important to at least label procedures with an ISO 14000 reference. This will allow the auditor to zero in on specifically what they need to audit, not extraneous material. Cross-referencing between the various documents will allow the important connection to be made. For example, cross-referencing will show the relationship between impacts, regulations and controls.

Health and Safety

One major question yet to be determined in terms of scope pertains to health and safety. Probably health and safety will not be required for certification. It would be in the best interest of the organization to design and implement both environmental and health and safety procedures, however. The standards definitely do not prohibit addressing health and safety and in fact encourage inclusion. These subject areas are so closely related that in many cases the same management act protects not only the environment but the health and safety of the employees as well. It is just a matter of time until health and safety are included, so they might as well be addressed now.

When

Standards are now being developed. Figure 3–1 shows a best-guess timeline for overall implementation and certification. This figure shows that internal preparation should start immediately. External certification will probably begin in the 1996 or 1997 time frame. It is in the organization's best interest to start preparing now, however. Preparation of missing procedures and systems will not be the time-consuming action. It is the

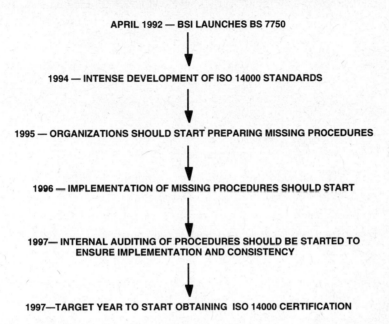

**PREDICTED TIMELINE FOR ISO 14000
IMPLEMENTATION AND CERTIFICATION**

APRIL 1992 — BSI LAUNCHES BS 7750

1994 — INTENSE DEVELOPMENT OF ISO 14000 STANDARDS

1995 — ORGANIZATIONS SHOULD START PREPARING MISSING PROCEDURES

1996 — IMPLEMENTATION OF MISSING PROCEDURES SHOULD START

1997— INTERNAL AUDITING OF PROCEDURES SHOULD BE STARTED TO
ENSURE IMPLEMENTATION AND CONSISTENCY

1997—TARGET YEAR TO START OBTAINING ISO 14000 CERTIFICATION

Figure 3–1

consistent implementation of the essential procedures that will take time. In fact, if the procedure is to be done by most of the employees in the organization, such as reuse of office paper, it may take years to implement.

Largely depending on customer demands, certain parts of the world may be faster than others in attaining certification. Europe will probably be first to require the certification mainly because they have been at the forefront of many ISO efforts in the past and because of their BS 7750, EMAS and eco-audit programs. Customers in the United States may be second to require certification of their suppliers. The Far East and rest of the world are expected to quickly follow.

SC1 Environmental Management Systems and SC2 Environmental Auditing will be the first to be required and the rest of the standards will be recommended or required later since they have been harder to prepare.

These later components will probably include environmental labeling, environmental performance, environmental aspects of product standards, design for environment, and life-cycle analysis. These concepts are all relatively new, at least in their practical application. Because of this, the subcommittees have not been able to move along as fast as SC1 and SC2.

Design for environment and life-cycle analysis requires organizations to introduce environmental considerations during the product design phase. For example, the ability for the product to be broken down into parts at the end of its useful life is important so that the components can be reused or recycled more easily. Design for environment is often very hard for many organizations to achieve because of intense competition and quality demands. As long as there is a process that continually reminds the design engineer to watch for environmental considerations and develop them when at all possible, certification can be obtained. This concept should be worked into the environmental management system as soon as possible; however, it will probably not be required for a few years.

Where

As with ISO 9000, implementation is occurring most heavily in Europe in terms of environmental management certification. The United States is also an area of great activity. Actually almost everywhere in the world is fair game and likely to be covered as well. As of the writing of this book approximately 90 different countries have endorsed ISO 14000 and will all surely be locations where ISO 14000 will be required by customers. Of the 90 countries, the most active include Australia, Austria, Belgium, Brazil, Canada, China, Denmark, Finland, France, Germany, Ireland, Italy, Japan, North and South Korea, Malaysia, the Netherlands, New Zealand, Norway, South Africa, Spain, Sweden, Switzerland, Thailand, Turkey, the United States and the United Kingdom.

Chapter 4

The Major ISO 14000 Subdivisions and Technical Committees

Introduction

It is only through the hard work of numerous individuals that environmental management standards are as far along as they are. Due to their extreme complexity these standards have been worked on by many groups and committees. An overview of some of the committees involved in ISO 14000 preparation is shown in Figure 4–1. As seen in this figure, there are committees working on ISO 14000 at the international, national and regional levels. The number of involved individuals worldwide is staggering. The primary committee is entitled TC207, and was established to incorporate the core environmental management systems into ISO. There are representatives from many countries on the committee. The United States representation is led by a technical advisory group (TAG) sponsored by the American National Standards Institute (ANSI). There are six primary subcommittees to TC207 and they include SC1–SC6. At this point in time only the specifications of SC1—Environmental Management Systems, will be the minimum requirements of certification. All the other subcommittees are preparing elements that will be recommended but not required (U.S. Sub Tag 1, 1995).

Environmental Management Systems (ISO 14001)

The SC1 subcommittee is really the heart of the entire certification effort. The environmental management system specification that they have produced is the only required

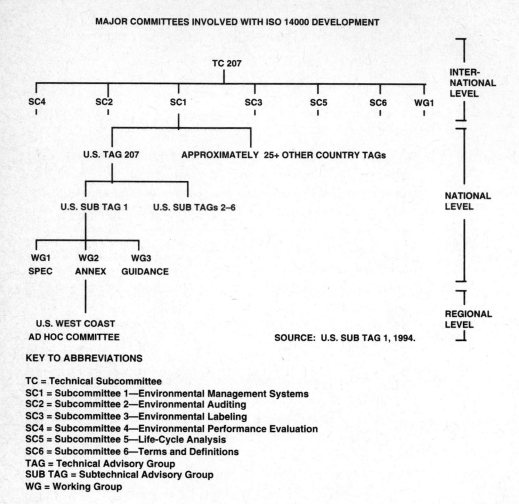

MAJOR COMMITTEES INVOLVED WITH ISO 14000 DEVELOPMENT

KEY TO ABBREVIATIONS

TC = Technical Subcommittee
SC1 = Subcommittee 1—Environmental Management Systems
SC2 = Subcommittee 2—Environmental Auditing
SC3 = Subcommittee 3—Environmental Labeling
SC4 = Subcommittee 4—Environmental Performance Evaluation
SC5 = Subcommittee 5—Life-Cycle Analysis
SC6 = Subcommittee 6—Terms and Definitions
TAG = Technical Advisory Group
SUB TAG = Subtechnical Advisory Group
WG = Working Group

Figure 4–1

specification at this time. The ISO 14001 specification is the most general and provides the overall framework for all the rest of the specifications (ISO 14001–ISO 14040). As of this date many different drafts of the SC1 documents have been released for comment.

The three SC1 documents of greatest importance include the specification, annexes and guidance document. The specification comprises the minimum requirements that must be met to achieve certification. The annexes present registration criteria, intent, links between ISO 9000 and ISO 14000, information for small and medium enterprises, history, discussion, interpretations, definitions and clarifications to the core elements. In reality the annexes are probably a semirequirement in practice. The guidance document presents best practices, newer concepts, options, and suggestions, such as risk assessments, that are beyond the certification requirements.

One annex of special importance is Annex D—a Special Guidance for Small and Medium Sized Enterprises (SME). A SME is defined as any organization up to 200 employees. The SME should prepare many of the ISO 14000 elements, such as objectives, targets, resources, guiding principles and the environmental manual, however not in as great depth (U.S. Sub Tag 1, 1995).

The United Kingdom leads SC1 and advocates following British Standard 7750 to a great degree. The U.S. representation to SC1 is sponsored by ANSI and has a slightly different approach than that specified in BS 7750. Both require, for example, identification and control procedures. One major difference, however, is that the United States recommends that impacts and regulations do not have to be disclosed to the public. The final ISO 14001 standard is targeted for release around December 1995.

Environmental Auditing (ISO 14010–14015)

Environmental auditing is closely related to the environmental management systems and will probably be the next implemented component of ISO 14001 certification; it is targeted for final release around October 1996. In general there seems to be better agreement between these first two subcommittees and their standards. As of this date several drafts of the SC2 specification have been released for comment.

Audits will concentrate on whether an organization has met the SC1 Environmental Management System (EMS) specifications and regulatory requirements. The audits should be objective, structured reviews to obtain and analyze information that will indicate the degree of conformance (Burhenn, 1994). In other words, have the systems been implemented and are they generating the results they were designed to produce?

The auditing can be done internally or by a third party. A third party audit costs more; however, in certain parts of the world third-party auditors are probably better able to handle the detailed audit criteria than internal auditors. The most compelling reason for using an external or third-party auditor is that they can be more subjective and will be regarded as such by others.

The Environmental Auditing subcommittee is led by the Dutch. Control of the overall environment management system is facilitated through auditing and correction of problems and periodic review of improvements. The five areas of concentration by this subcommittee are audit principles, audit procedures, qualification of auditors, site assessments, and environmental investigations. Presently SC2 are recommendations, not requirements (*Environment Today*, 1993).

Environmental Labeling (ISO 14020–14024)

This subcommittee deals with all types of environmental claims, advertising and marketing. In addition to labels, TV and radio advertising is covered. Other programs and regulations, such as the "Green Dot" in Germany may be impacted by SC3 activity. One goal of this subcommittee is to bring some consistency and uniformity to environmental labeling, which differs significantly from country to country. For example, there is little similarity in the Blue Angel, Green Cross, Green Dot and the numerous other labeling systems.

The Environmental Labeling Subcommittee is led by Australia. This seemingly small topic has been assigned to a separate subcommittee since there are so many conflicting labeling programs in existence in different countries. Unfortunately many of the labels have and still are given to organizations who truly have not earned them. Unsupported claims in many cases have resulted in the granting of labels to many undeserving applicants (Morrison and Foerster, 1994).

The labeling recommendations will cover three types of labels ranging from general to specific. The first type would indicate that the product is environmentally friendly. The second type concerns a specific manufacture's claim, such as recyclable. The last deals with environmental effects of the product (Burhenn, 1994). No matter what the level or type, the most important consideration are fairness and consistency in labeling that can be recognized around the world.

Environmental Performance Evaluation (ISO 14031)

The United States leads the Environmental Performance Evaluation (EPE) Subcommittee. As the name implies, this subcommittee specifies the degree of success that is expected in various environmental management programs. This concept also includes periodic improvements in performance as a component of certification. Unfortunately at this time EPE is still rather subjective in nature.

There is a difference between EPE and auditing. EPE applies continuously over time whereas auditing applies to one moment in time. EPE is the ongoing measurement of how well an organization is performing and improving. Numerous indicators are to be used including environmental, operational and others (Burhenn, 1994). EPE promotes the concept that a quick audit and fix are not the way to go. Rather, it is the long-term environmental performance actions and their measurement that are really important for long-lasting environmental improvement. EPE is a process used to measure, analyze and assess an organization's environmental performance.

The EPE concept uses an environmental performance indicator (EPI) as a central component. EPIs measure how well the organization is doing against set criteria. The EPIs are the actual items assessed and tracked (Morrison and Foerster, 1994). The EPIs are very similar and in many cases identical to the measurable targets that an organization sets up. An EPI can also be a more detailed level of assessment or breakdown of the targets, depending on the sophistication of the organization.

Life-Cycle Analysis (ISO 14041–14044)

There is less agreement about the direction and standards for the Life-Cycle Analysis (LCA) subcommittee and the remaining subcommittees presented in this chapter. Because of this and the complexity of the subjects, it is improbable that these will be added to ISO certification in the near future. It is still good to keep these concepts in mind because they can have a tremendous impact especially on product design and marketing. For example, time to market can be slowed down due to LCA.

Germany and France are leading the Life-Cycle Analysis Subcommittee. Life-cycle analysis and design for environment are slightly different concepts. In general, however, both consider environmental impacts during the entire life of the product. This includes the conceptual design phase of a product, raw materials, operational impacts, recycling or disposal of the product. For example, the ability to disassemble a product at the end of its life must be considered even before the first product rolls off the production line. The four major subject areas being considered include policies and practices, general principles, life-cycle inventories and methods of impact and improvement (*Environment Today,* 1993).

LCA is a very complicated process with many qualitative and quantitative steps. In general, however, there are three major categories of actions. The first type is the impact inventories of product (and raw material). Once all impacts have been identified an analysis occurs. The analysis will be different for various types of impacts. For example, a risk assessment might be used for certain types of impacts originating from certain raw materials. The final step or category of action would be an improvement analysis or a determination of how the impacts can be minimized (Burhenn, 1994).

Terms and Definitions (SC6)

The Terms and Definitions Subcommittee is led by Norway. Their charter is to coordinate the use of comprehensive and standard terminology among the subcommittees. This will be difficult considering the fact that many different countries along with their unique languages and legal systems are represented on the subcommittees. Having a common ground or language is extremely important in terms of a worldwide standard. Countries must first agree on definitions or language before they can reach agreement on the more qualitative matters. In hindsight it probably would have been better to designate this group SC1 and had it do its work first.

Environmental Aspects of Product Standards (ISO 14060)

The Environmental Aspects of Product Standards Working Group is led by Germany. Many of the systems required or recommended by other subcommittees are referenced in these standards (Morrison and Foerster, 1994). The primary focus of this working group (WG) is to provide guidance for standard writers. Criteria will be developed for the writers in terms of environmental impacts. The details of this subject are not yet clearly defined and will therefore probably be delayed for some time. It will be extremely difficult and time consuming to come up with standards that can be applied to any country in the world.

Possible Future Working Groups or Work Items

Since environmental management is a relatively new field there are unlimited possibilities for future working groups. Many of these topics are academic; however some will have significant environmental and business impact. Environmental site assessments and product chemical risk management have both been mentioned as possible future subjects for committee or work-group action.

Chapter 5
General Steps to Achieve Certification

Introduction

All components of the environmental management system should be coordinated with other important functions of the organization, especially at the policy level. For example, the policies, objectives and targets of the finance, operations and safety departments must be considered and, if possible, be compatible with those of the environmental department.

If an organization is already ISO 9000 certified or at least close to it, considerable time can be saved. For example, the following materials already prepared for ISO 9000 certification can be used for ISO 14000 with only minor change: organization and personnel procedures, records and control of documents, audits and reviews. Many other sections from ISO 9000 documents can be used as a starting point; however, considerable modification would need to be done, for example on the procurement section, to bring them up to ISO 14000 standards (Rothery, 1993).

Most organizations are already in the process of environmental management. Therefore the starting point in the certification process will be different for each organization depending on how long they have been in operation and how detailed their present environmental system is. Later chapters in this book present greater detail for most of the following steps.

Initial Assessment and Definition of Purpose

Before an organization begins massive design and implementation efforts, an initial assessment should be done. This will help determine where the most critical needs exist for new environmental management systems. For example, has an effective policy been established by executive management? If most major systems are in place then the initial assessment will suggest where upgrades are needed. A formal initial assessment document is not required for certification. The assessment should identify documents, actions and procedures that are required for certification, such as a policy statement, the management system, planning, operations, personnel, training and goals.

A definition of purpose should be made in association with the initial assessment. The purpose could be to better protect the environment, to become ISO 14001 certified, to become more cost effective, to improve community relations, to improve market appeal and numerous other purposes.

To be successful, the entire certification effort should be summarized and presented in the initial assessment and approved by top management of the organization. The initial assessment should be presented to top management in a fashion that will attract their attention. This presentation may come from the organization's environmental manager or an outside consultant. The initial approach can be either verbal or written, depending upon the situation. For example, it may be suggested at a routine meeting of company management.

A preferable method of introduction over casual mention during a meeting is by an internal letter to one or two of the executive managers. Figure 5–1 is an example of a letter or memo that could be used. It is important to make the correspondence clear and brief. A telephone call or a letter from a consultant briefly suggesting the need for the certification and offering to outline further details and benefits are other possible introductory ways.

Whoever approaches management initially must, without fail, be thoroughly familiar with ISO 14000 and how these standards will apply to the specific organization. They should be prepared to briefly outline at the first meeting the background of the standards and their impact on the organization, consumers and the public. The individual's knowledge and presentation at this meeting will determine the degree of acceptance of

TO: CEO

FROM:

SUBJECT: ISO 14001 certification

I need to bring to your attention that there is currently a monumental effort by countries throughout the world to establish a universal standard for environmental management. These standards will affect all phases of our operation. Organizations who meet and achieve these standards applicable to their areas of endeavor will be certified by independent ISO auditors, as was the case with ISO 9000. Certification is not mandated by any country or political organization but effectively becomes necessary if one is to remain competitive in business.

We have followed the development of these standards and are in a position to be of service to you to obtain certification. I suggest a preliminary meeting with you to discuss this in greater detail.

Figure 5–1 Example of an ISO 14000 Concept Introduction Letter.

the certification process by management. The advocate must be prepared to answer searching questions in all areas relating to certification. An outline to obtain certification could also be presented at this time.

The initial assessment can be conducted by a consultant or by in-house personnel. Whoever is responsible for the assessment must view the work and environmental effects created by the company from an entirely neutral or independent view. If not, a genuinely effective policy based upon this assessment will not be developed. The resulting program will not attain the necessary results and certification will be delayed or never obtained. One can say, therefore, that the future of the organization may depend upon the thoroughness, accuracy and integrity of the initial assessment. All initial assessment work with results and conclusions must be documented for reference during certification procedures.

The initial assessment should include the following:

- *Copies of ISO 14000 Standards*—At least the standards generated by SC1 or the EMS standards should be attached along with the annexes and guidance documents for SC1. The standards prepared by the other subcommittees, which are not required, would not have to be attached.

■ *A Listing of Major Applicable Regulations*—A listing of applicable regulations and the sections of the most important regulations should be assembled. If this is too great of a volume of paper, then it might be more meaningful to attach regulation summaries, guidance documents or regulatory summary charts.

■ *A Listing of Major Impacts of the Operation*—Considerable attention to detail is especially important for this step. All impacts, no matter how small, should be identified. Quantification and elimination of insignificant impacts will occur later. Impacts should be identified that are associated with the site/operation, raw materials, vendors, product and/or service.

■ *Current Environmental Controls*—A listing of current actions should include their effectiveness, completeness, staffing, funding and present top management support. This would also include procedures in place, use of consultants and any major environmental control system currently operating.

■ *Additional Activities Needed and Areas to Be Covered*—At this point in time a best guess should be made of what systems should be added to protect the environment. These may later become recommendations made to management to improve environmental controls and/or allow certification. As the assessment and planning continues, more systems will become obvious.

■ *Estimated Cost and Benefits*—When all these recommendations are made to management, the first questions asked will probably pertain to the costs and benefits. If they are at least estimated roughly at this point, funding activities can be started earlier.

All of the remaining sections in this chapter are discussed in greater detail later in this book. They are presented in summary here so that the reader can quickly see the entire process from beginning to end.

Policy Preparation

A comprehensive environmental policy statement that will cover all of the employees in the organization should be prepared. The policy should address impacts and regulations in a broad sense. It should also be supported by senior management and communicated to all employees and interested public. Details concerning policy preparation are presented in Chapter 6. The policy preparation should start very early in the process and be continually upgraded.

Policy preparation needs to be done very early in the process since upper management approval is key to success. Even if the policy is only roughed out at this point, it will at least provide overall direction to the process when it is especially needed. A rough draft of the policy should be widely circulated for upgrades and input from as many employees as physically possible. This will increase acceptance to the maximum extent possible.

Obtain Up-front Resources

Certain resources will need to be obtained early in the process in order to complete the rest of the steps. Financial resources are the first that must be made available. Organization resources, such as personnel, may also have to be established, if they are not already in place. Once this occurs training resources can be identified and provided. Purchase of supplies and other support should also occur. Additional detail is presented in Chapter 7.

Prepare Procedures for Identification of Impacts and Requirements of Others

Procedures for identification, assembly, and analysis of impacts and regulations into the organization's systems are needed. Even though they are not required for certification, it is a good idea to actually obtain the regulations and impacts and assemble them into one or two binders. This step is essential and allows for the meaningful design of environmental management procedures and controls. This topic is more fully discussed in Chapter 8 and Chapter 9.

Objectives and Targets

Objectives and targets should next be prepared in order to achieve the policy statement. The impacts and regulations previously identified should also be considered when the objectives and targets are prepared. The objectives would include statements such as establishment of a waste minimization program. Targets would be specified for each objective and present numerical goals, such as 10 tons of acetone waste recycled in 1996. As with the policy statement, the objective and target preparation should start early and be continually upgraded. Additional detail is shown in Chapter 10.

Utilization of Existing Documents and Resources

If documents already exist for successful programs that address some of the ISO 14000 considerations, they should be utilized. These could include most of the good environmental management procedures and many of the quality systems already in place. For example many ISO 9000 documents and systems, such as training, could be used as is or with very little adaptation since there are many common links or elements. Appendix B of the actual SC1 specification lists the documents that are common to both ISO 14000 and ISO 9000.

Preparation of New Operating Procedures and Action Plans

Once the above steps have been completed it is time to prepare the new procedures that are missing. Far too often many environmental management systems are composed of unwritten procedures and standards. This usually leads to confusion, lack of direction and negative environmental impacts. Even if an individual plans to do an environmental control action only one time, it is still good to put it in writing so that questions can be answered when they come up about what was done.

An environmental management manual(s) should be set up if one does not already exist to contain all the different procedures and standards. It should also contain a copy of the company policy. A procedure or binder that deals with regulations and one that presents the procedure for identifying and dealing with impacts are additional examples. If a very rough draft or initial collection of existing materials is done quickly, the task will not seem as overwhelming. Chapters 11 to 19 present additional information concerning procedures and action plans.

Implementation of Programs

Now that the paperwork is done, it is time to start the real action. Far too often some organizations only write and talk about environmental protection. The environmental management systems need to be implemented to actually help the environment. This may seem obvious; however, in real life administrative delays, apathy, and inadequate personnel and training can prevent environmental protection and improvement from happening. A good top management policy, energetically pursued will assure success. Knowing that continued operation depends upon innovative environmental protection, good management will create the necessary authority and funding for an effective program. The bottom line is that certification will require implementation in addition to the preparation of procedures and documents.

Ongoing Auditing, Management Reviews, Correction and Follow-up

By continually auditing or reviewing the progress an organization is making in environmental management, it is possible to suggest meaningful correction and follow-up. This field is changing so rapidly that adjustments need to be made almost daily. Once the audits are completed, the corrective actions need to be made promptly. Liability problems develop if the file shows correction was needed but not done. Chapters 14, 15 and

16 provide details concerning these subjects. The audits, reviews, correction and follow-up will result in continuous improvement of the environmental management system. It is always possible to improve the quality of the environmental controls with a net positive impact on the organization and the living things within and around it.

Internal ISO Standard Audit

When an organization feels it is close to completing the items noted above, it is a good idea to do an internal audit. This type of a practice audit will help identify last minute corrections that still need to be done. The audit is only valuable if the internal auditors are trained to be critical of their own organization, which is sometimes hard to do. A properly performed audit by well-trained internal auditors who are given sufficient latitude to perform their jobs will save the organization time and money in the long run. It will be far cheaper for an internal team to identify and facilitate correction of as many problems as possible than to hire external auditors.

Outside Auditor Audit

The actual audit for certification purposes is usually done by outside independent auditors. More credibility is usually given to these third-party audits because they are felt to be less subjective. On the other hand the auditors may not be familiar with the particular industry being audited. If this is the case it is a good idea to start a positive relationship with the auditors by providing some up-front technology education. Overall, however, the auditors must be qualified to do environmental management auditing and approved by ISO before they can give certifications. These individuals are highly trained in auditing and pride themselves in being tough. If ISO 9000 is any indication it would not be surprising to see a 50 percent certification failure rate for first-time audits.

Certification

Based upon ISO 9000, if the organization passes most components of the audit, the certification will be awarded. This would apply only if the failed components are not considered major deficiencies. Failure requires corrective actions and reassessment within a specified period of time.

Certification can occur in three different ways. If it is done by an outside independent auditing firm that is approved by ISO, the certification will carry the most weight. A second-party certification occurs when it involves suppliers under contract. In this case the audit could be done by the organization that uses the supplier. Self-certification obviously carries the least weight; however, it is better than no certification at all. No matter which type of certification method is selected, it is at least a proactive step in the right direction.

TC207 is promoting integration and certification of ISO 14000 with ISO 9000 via a coordinating committee (TC176/TC207). Integration is recommended by some since they are related and could complement one another in certain aspects. Integration, however, could threaten certification of both if one is not up to the standard. Whether they are officially combined or not, they must at least be compatible.

Continual Improvement

By doing routine internal audits and monitoring, it will become evident that the policy, objectives, targets and plans will have to be modified. Frequently upgrading the entire system will keep it cost-effective and impacts will be reduced to the maximum extent possible. Continual improvement is not really a last step. It is an integral part of every step in environmental management whether it is mentioned or not.

Part II  Initial Environmental Elements and Components Needed

Now that the background information has been presented we can move into the details of the actual elements or components of an environmental management system that should be set up first. The chapters in Part II provide information an organization can use to prepare their own site-specific elements. These must be in place in order to obtain an ISO 14001 certification and are just plain good environmental management even if a certification is not desired.

The chapters in Part II are presented in the order that the author recommends an organization follow in establishing the elements. There is a logic to the progression of first setting up a policy, followed by getting the environmental employees in place and organized, assessing environmental impacts and regulatory requirements and then establishing environmental objectives and targets. This recommended order of preparing the initial environmental elements of a management system is especially important for the start-up organization with little in place. Organizations that have been around for awhile may find many of these elements in place, at least in part. If this is the case they should not throw them all out and start from scratch. Elements and components of environmental management systems that are working should be saved. The well-established organizations should concentrate instead on the missing pieces and continually upgrade the established elements.

Chapter 6
Environmental Policy

Introduction

We now turn our attention to details of the actual components of the environmental management plan. These components or programs are presented roughly in the order in which they should be developed. First, following the preassessment report is the policy statement.

The policy statement is a declaration, signed by the top officer(s) of the organization, that environmental protection is a priority. At a minimum the organization's president should sign the policy statement. It is also a good idea for the senior vice presidents to sign as well since their support is also essential. Without this show of commitment from upper management the rest of the organization will not be dedicated to environmental management.

Top management needs to show enthusiastic support of the policy statement in several ways. In addition to signing the statement, showing support by providing adequate funding is of extreme importance. Lacking this, the company, or at least the environmental management portion, will fail. Support of the policy statement can also be shown by senior management's saying so in memos, videos, meetings and actions. For example, if the president of the company is seen using the back of a used sheet of paper, what better message is there that he or she really cares about the environment?

Key Components of the Environmental Policy Statement

All policy statements need to have certain characteristics if they are to be successful. First, they should be relevant and straightforward. The statement should relay that protection of the environment is a top organization priority. A commitment to continued improvement of environmental performance and compliance with laws and regulations are key. The statement should clearly specify which organizational activities (hopefully all of them) are covered by the statement. The statement should be a natural jumping-off point for setting environmental objectives and targets. It should also provide a framework for assessing progress made with the targets and objectives that are oriented toward minimizing environmental impacts. In other words, the policy statement provides an environmental purpose and set of values for the organization to follow.

Communication, Promotion and Support of the Policy Statement

No matter how good a policy statement is, it will be totally ineffective if the commitment it contains is not communicated, available, promoted and supported. At a minimum it should be available to all employees in the organization. It is best if each employee can see the statement during at least part of their workday. The importance of communicating the statement can't be overstated.

There are various ways to communicate the policy statement and repeated exposure is key to all of them. Policy statements can be posted in conspicuous places, printed in organization newsletters, sent to employees as a desk drop and communicated in numerous other ways. After a period of time the communication should be repeated, as a reminder.

The policy should be made available to the public, as well as employees. Employees and the public who can see a written commitment by the company to effectively control environmental impacts in all areas of its operation will show an increase in morale and faith in the organization. Whenever a copy is requested, whether by an employee or someone outside the organization, it should be promptly provided.

In summary, there must be top-down support of the policy statement. This is shown not only by signing the statement but by involvement and promotion of the statement at all levels. For example, if upper management continually relays to the employees that they feel the environment is important, the entire organization will act accordingly. This basic message and day-to-day actions at all levels of management will show support of the policy statement.

Types of Policy Statements

Environmental policy statements range from the very general to specific. All organizations should have at least one general statement that will cover the major environmental issues. The issues do not have to be named specifically; however they should also not be excluded in concept. Some organizations combine health and safety into the general environmental policy statement. Since the subjects are so closely related this combination or marriage of disciplines at the policy-statement level is a good idea. Even if different teams of employees are handling environmental policy and health and safety, they will be more apt to work together as a team for the health of the employees and the environment when they are all included in the policy and it is clear to them.

There are numerous possibilities for specific environmental, health and safety policy statements. Depending on the environmental problems and opportunities facing an organization it might also be good to have one or two specific policy statements. For example, waste minimization is commonly a second or subpolicy statement because of the great importance of this "way of the future." Employee injury reduction can also become a stand-alone or subpolicy statement.

Examples

General Environmental Policy Statement

The following example illustrates one possibility for the lead-in for a general type of environmental policy statement. The organization name would be inserted, along with the signature of the president and senior vice presidents:

> Senior management believes that no aspect of operation is of greater importance than environmental protection. This organization's policy is therefore to provide and maintain the most effective environmental control procedures, to follow all applicable legal requirements and to do everything that is feasible to protect the environment.

Some organization-specific elements should also be added to make the above policy statement more meaningful. For example, if multiple locations are involved, it is important to state that there will be on-site environmental personnel to handle environmental issues.

Specific Policy Statement

An example of a specific type of environmental policy statement, waste minimization is given below. Again the name of the organization and the signature of the top officer(s) would be inserted:

> Waste minimization is to be practiced by all employees, to the maximum extent possible at all our facilities. This is of great value to the environment, to the safety of our employees and continued success of our operation. Reduction of chemicals and wastes and purchase of recycled products are priorities for all employees. These efforts should apply to both hazardous waste and to the nonhazardous wastes as well, such as paper, plastic, glass, aluminum and cardboard. All employees must work together to aggressively seek and implement waste minimization measures. These measures include material, energy, and water reduction, recycling, reuse and treatment and all help protect the environment and employee safety.

Revisions/Continual Improvement

Since the environment as well as the organization changes, it is essential to consider adjusting and improving the policy statement from time to time. If it is prepared correctly, however, it should not require adjustment every year. In fact the policy statement should provide some stability since most other elements in the EMS need constant adjustment.

Chapter 7
Organization and Personnel Issues

Introduction

In the last chapter the policy statement that must be endorsed by the highest individual in the organization and general environmental resources to handle problems was discussed. This topic is continued in this chapter which deals with detailed organizational issues for all employees including the top individual.

Unfortunately organizational issues are often overlooked when it comes to environmental management. The attention is usually on technical aspects of environmental control, at the expense of human and organizational considerations. This is especially detrimental when one considers the fact that environmental issues, such as hazardous waste can be highly emotional topics. Most environmental management references direct very little attention to the organizational issues which can be very sensitive.

In the past many organizations have considered and treated the environmental department as a stepchild or a necessary evil. As a result, some of their programs are poorly understood, meagerly funded, undermanned and with little or no recognition, support or direction from top management.

The incorporation of environmental management standards into ISO will establish environmental management as one of the critical operations, equal in importance to many other departments. Although many will believe this status is not warranted, except the customer, as time passes and the benefits of an aggressive environmental program are realized, such opinions will change rapidly. In addition, under the new standards, environmental certification must be established and maintained if the products are to be successfully marketed.

Employment of innovative environmental individuals will prove to be of great value in the future. Over the long run, environmental costs will not seriously affect the competitive balance between companies because each must meet the same standards. Innovation and creativity by employees to satisfy the standards will help the organization obtain a competitive edge.

Certification could be coordinated by the environmental department or a joint effort by the environmental and quality departments. Factors influencing the decision concerning who should lead the effort would be technical experience, size of the organization, amount of work needed to obtain certification and the present degree of functional overlap between the two departments. Whichever department is chosen to coordinate should work closely with the other department.

General Organizational Requirements and Resources Needed for Certification

Responsibility, Authority and Job Title

At least one management representative must be given the authority to ensure implementation of the environmental management systems. This individual must be high enough in the organization to authorize resources when they are needed to protect the environment. The person must have the full support of the top officer in the organization to be effective in ensuring that the environmental management policy statement is followed. This individual should report the status of the policy, objective and targets.

For the rest of the employees in the organization, authorities and responsibilities must be clearly spelled out, not only for certification but for effective management. This is especially important in environmental management since there are so many "gray area" actions that will need to be performed. The employee will continually be wondering whether they should be involved in a particular task or not. If each employee sees their responsibilities in written detail, they can ask questions if they are uncertain of what is expected of them. This will reduce doubts in employees' minds regarding their specific areas of responsibility and will help establish a sense of purpose in the employees. They must know when they have authority to make decisions and when they do not.

Examples of some job titles and associated job descriptions are shown in Appendix A. Note that the responsibilities are spelled out in detail. In addition to those shown, there are numerous other environmentally related job description possibilities. For example, the environmental engineer job description could actually be broken down into associate, junior or senior level. There are numerous other possibilities for environmental type job titles, each with their own unique set of responsibilities. For example, the following list presents a few possibilities:

- Chemist
- Environmental Laboratory Manager
- Epidemiological Manager
- Environmental Training Specialist
- Hazardous Materials Specialist
- Manager of Hazardous Materials Management
- Environmental Protection Engineer
- Air Quality Engineer
- Safety Engineer
- Health Physics Technician
- Risk Analyst
- Manager of Risk Assessments
- Toxicologist
- Industrial Hygienist
- Hydrogeologist
- Regulatory Compliance Specialist

- Environmental Technician

- Environmental Scientist

- Environmental, Health and Safety Manager

Special situations may require the use of a "delegation of authority" form. If the supervisor is going to be away and wants the employee to handle nonroutine situations that do not fit into their job description, the delegation of authority form should be used.

Number of Personnel Needed

An "adequate" number of employees must be involved in environmental management activities. The interpretation of the word adequate can vary widely. There is flexibility in the number of full-time and part-time employees and contractors. The important thing is that the issues are being addressed by someone on behalf of the organization. It would be impossible to give a magic number or formula of environmental personnel required since each organization and location is different. One way to sense whether the number is even close to realistic is to determine whether at least half of the full-time environmental employees are working on proactive environmental projects vs. everyone putting out environmental "fires." If almost everyone is involved in short-term projects of a crisis nature, more employees are needed. One possible way to assure yourself and the ISO auditor that an adequate number of environmental employees are present follows. Count the

1. Number of personnel involved with cleanups

2. Number of personnel involved in litigation

3. Number of personnel involved in fine resolution

4. Number of personnel involved in waste minimization

5. Number of personnel involved in vendor control

6. Number of personnel involved in environmental training

7. Number of personnel involved in other long-range environmental management

The total number of personnel (employees and contractors) involved with items 1 to 3 should not be greater than 50 percent of the total number involved with items 1 to 7. The goal is to redirect the effort from putting out environmental "fires" to doing proactive environmental actions.

In 1995, an informal survey was done of 12 different international electronic manufacturing companies to determine the average ratio of environmental, health and safety (EHS) personnel to the general employee population. Since some individuals were performing both environmental and health/safety functions, it was impossible to get numbers for environmental professionals only.

The ratio varied dramatically depending upon a company's level of general environmental awareness and proactivity, its past compliance history, its location and the amounts of chemicals and hazardous wastes its processes involve. Unfortunately, the past compliance history, such as fines, cleanups and litigation, was one of the major factors behind higher numbers of EHS employees.

The data showed that the company with the highest ratio of EHS employees per general population had 1 EHS for every 14 employees. The company with the lowest ratio had 1 EHS for every 1,450 employees. The average of all twelve companies was 1 EHS for every 170 employees.

It would not be wise to use this ratio as the absolute justification for increasing staff since only 12 companies were considered. The ratio does, however, provide a starting point for discussions concerning the proper staffing levels. It would then be factored up or down depending on the organization's specific situation.

Personnel Caliber or Quality Requirements

The nature of environmental management requires that the employees involved be motivated and of the highest caliber. They must be competent, trained and have experience if they are to make meaningful environmental improvements and obtain certification in a cost-effective manner. Even if they may seem obvious, some suggestions for assuring the caliber of environmental employees follows:

- *Degree*—Minimum of a B.S. degree in environmental management, chemistry, ecology, engineering, waste management or other related subject.

- *Experience*—An adequate amount of experience, depending on the amount of supervision. At least five or more years of experience should be required unless close supervision is maintained.

- *Inquisitive*—An inquiring nature with an innovative mind. This is especially important since the environmental management field is relatively new and changing rapidly.

- *Registrations*—Working toward or have registrations and certifications such as the professional engineer, certified industrial hygienist (CIH), registered environmental manager, registered environmental assessor.

- *Good Personal References*—References should be good, not only those provided by the applicant, but from other nonlisted acquaintances as well.

- *No Criminal Record*—Lack of a criminal record may seem like an obvious requirement, however, it would surprise most readers how many job applicants have criminal records over and above traffic and DUI convictions.

Assuming that the employees meet these criteria, it might seem unnecessary to also have a "Code of Conduct"; however, it doesn't hurt, especially since one is required for certification. An example of one possible Code of Conduct for the environmental staff as well as all other employees in the organization is shown on Table 7–1. The Code of Conduct should be general and not conflict with the specific responsibilities which are spelled out in detail in each procedure.

Organization Structure

Environmental departments can be organized in almost any way in terms of organizational structure. Whatever it takes to get the job done is acceptable. In many situations, however, the traditional "mechanistic" or

Table 7–1 Code of Conduct and General Responsibilities (Example)

All individuals in this organization are responsible for helping to protect the environment and employee health and safety. Your conduct must clearly demonstrate that you are meeting this important responsibility. More specifically:

■ Every individual must help protect the environment during the performance of their daily job. All actions that add air, water or solid waste pollution to the environment must be minimized.

■ Every employee is responsible for making sure their work area is safe and they perform their work safely. Each employee is also responsible for minimizing waste and handling it properly.

■ Every manager and supervisor is responsible for inspecting their area for environmental, health and safety hazards and arranging for training in these subjects.

■ The corporate and/or field environmental personnel must track regulations, implement compliance programs and assist in reducing impact to the environment. Their conduct must be of the highest quality in terms of environmental protection so that they are promoting the right message.

■ Every officer is responsible for ensuring that environmental, health and safety policies and procedures are followed in their organizations and for providing the resources needed to meet them.

pyramid-type organization structure is not well suited to quick resolution of an environmental crisis or implementation of innovative and proactive programs. The "organism" or flatter structure is better suited for these situations, which are usually handled by experienced, professional employees.

Financial Resources

It is hard to provide a formula for financial resources as it is for number of employees needed. Money will have to be spent to either proactively minimize environmental impacts or to clean up environmental problems once they occur. Obviously, proactive management is going to cost the organization considerably less over the long term than to perform a massive cleanup of soil and ground water later.

Figure 7–1 illustrates one possible way to present financial resources. This type of presentation could be used to obtain budget approval from senior management or to illustrate the financial aspects of environmental, health and safety work to the accounting department. The presentation could help illustrate either past expenses or future environmental expenses, which are extremely hard to predict.

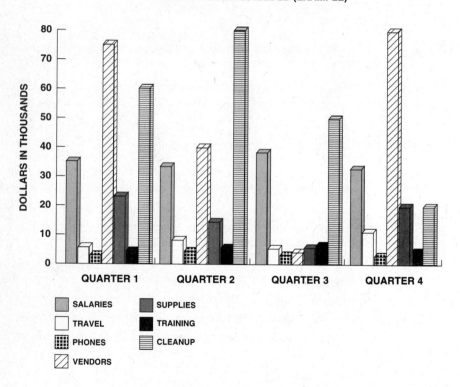

Figure 7–1

One possible way to ensure that the proper financial resources have been earmarked in the budget and to satisfy the ISO auditor is to show the number of dollars in the present budget for

■ *Environmental Personnel Salaries*—Salaries should be listed for not only the individuals doing full-time environmental work but also for those who do other work in addition to environmental.

- *Environmental Fees, Taxes and Penalties*—Permit and hazardous waste fees and taxes should be identified. It may be impossible to identify regulatory penalties; however, it should be kept in mind that they have been high enough to bankrupt a few organizations.

- *Environmental Supplies*—Dollars should be identified in this category to cover paper, pencils, forms, spill cleanup materials, floppies, sample containers, and so forth.

- *Proactive Environmental Projects*—Money will be required to set up proactive systems. For example, most waste minimization programs would be considered proactive and have an associated up-front cost. Purchase of recycling containers is an example.

- *Environmental Training and Development*—Environmental professionals need ongoing training and development, no matter how qualified they are when they start the job. Many of these training programs and classes are expensive and involve considerable time. For example HAZWOPER training lasts for 24 to 40 hours and is required for all employees who treat hazardous waste and serve on Emergency Response Teams (ERTs).

- *Travel for Environmental Issues*—Usually the employee will have to travel to where the class is given, to the agency office or the specific location of the emission or impact. These travel costs should be identified, even if some are hard to predict.

- *Environmental Vendors*—Due to the highly specialized nature of the field it is impossible for any one individual to be able to handle every environmental issue. Therefore vendors will be needed from time to time and usually at considerable cost. Hazardous waste vendors involved with transportation, treatment and disposal will also need to be identified along with their cost.

- *Cleanups*—Possibly the highest cost and the hardest to predict will be cleanup costs. If only soil is involved, the cost may not be as high. If, on the other hand, ground water is also contaminated the cleanup costs can be many million dollars.

Accounting codes should be set up so that all these categories can be tracked. This will help in forecasting budgets, auditing performance and answering agency questions. In addition, upper management needs to know the true cost of environmental management. If even half the dollars were tracked, it would become obvious to all employees, including the accountants, that environmental management is a very important budget item.

Information Resources

For the environmental individuals to be able to do their jobs efficiently, they must have access to various information resources. These vary greatly depending on the product and area. However, at a minimum, a chemical database and a legislative database should be available. Other types of resources that should be considered include computer listings of trade associations, agencies and technical literature databases.

The availability of adequate informational resources can't be stressed enough. It is impossible for the environmental manager in an organization to know all the answers in such a broad field. Therefore they must have ready sources of information for when the questions arise.

Training

Training is absolutely essential for environmental management due to the complex nature of the subject. The training is needed for not only the environmental staff but for all the other employees in the organization and certain contractors as well. All employees must be made aware of their impact on the environment during the performance of their jobs and ways to minimize the impact. Degree and scope of training will depend upon specific job functions. Whatever is needed to allow employees to help protect the environment and minimize risk to their safety and comply with the regulations should be provided. The entire training program must be continually reevaluated and updated, since the regulations and business requirements frequently change.

Identification of Training Needs

A procedure is needed to identify training needs of the individual. The scope and depth of training should be appropriate. The procedure does not have to be involved but should at least cover the following basic elements:

1. *Identification*—The first step is to identify nonhazardous waste, chemicals and hazardous wastes that employees might encounter. Emissions that employees might encounter should also be listed.

2. *Regulatory Requirements*—Laws that require training for the chemicals and hazardous wastes identified (Occupational Safety and Health Act, Resource Conservation and Recovery Act, Hazardous Material Transportation Act, etc.) would next be assessed.

3. *Employee Assessment*—The knowledge, skills and abilities of the employee should be assessed in reference to requirements 1 and 2.

4. *Match Process*—A match of employees (by name or job type) with the required training is performed next.

An example of how the training identification might look is presented in the Table 7–2 training matrix. This table shows different training classes across the top and nine categories of employee along the side. A "•" is placed in the rows to indicate that the training is required for that particular employee type. This table also shows the recommended length of the training.

Providing the Training

It is important to convey to employees during the training why ISO 14000 matters in their personal lives as well as to the organization. For example, if the trainer relays that implementation of ISO 14000 will help keep the earth livable for them and their children, they may develop a vested interest and pay attention during the class.

Adequate resources must be available to provide the training identified in a timely manner. Some training, such as first aid and CPR, must be provided by certified trainers. As with corrective actions identified in an audit, you don't want to just identify and not follow through. The training needs to be provided upon hiring and reinforced by an annual refresher as well.

Table 7–2 Environmental Training Matrix

	CLASS NUMBER								
	1	**2**	**3**	**4**	**5**	**6**	**7**	**8**	**9**
LENGTH OF CLASS IN HOURS	2	1	8	20	1	4	4	8	2
POPULATION All employees		•			•	•			
Chemical handlers	•	•			•	•			•
Engineers		•			•	•			
ERT Members	•	•	•	•	•	•			
Maintenance	•	•			•	•			
Supervisors		•			•	•	•		
Hazardous waste handlers	•	•	•	•	•	•		•	•
Warehouse	•	•				•	•		•
Environmental staff	•	•	•	•	•	•	•	•	•

KEY:
• = Class required
Class 1 = Chemical Handling
Class 2 = Hazard Communication
Class 3 = Hazardous Waste Handling/Management
Class 4 = Emergency Response
Class 5 = Job-Specific Training
Class 6 = ISO 14000 Conformance/Environmental Awareness
Class 7 = Supervisor Environmental and Health Training
Class 8 = Wastewater Treatment
Class 9 = Hazardous Material Transportation

The environmental training is usually provided by a variety of resources, depending on size and type of employee population. For example, if there is a large number of employees, it may be more cost-effective to have an internal training department provide much of the training. In some smaller organizations the employees may have to be coached or trained by the site environmental manager. The corporate environmental department itself should plan to provide certain types of training depending on their own workload. Consultants often provide on-site or "suit-case" training or off-site training. HAZWOPER is an example of training usually provided by consultants. Whatever the mix of training resources used, which is at the discretion of the organization, it could be noted on a matrix with footnotes such as:

1 = Corporate Environmental Department Provided

2 = Training Department Provided

3 = Consultant Provided

4 = Site Environmental Manager Provided

5 = School, Seminar, Video, and so forth.

Tracking the Training Hours and Other Employee Data

To show compliance with the laws mentioned above and to obtain ISO certification, the training hours and other employee data need to be collected and tracked. It is not enough to just keep an attendee list for the file. Total training hours should be calculated and reported to management. Hours per employee also need to be determined to ensure that goals are met. It is also a good idea to develop a computer database that matches employee name and job code with required training. A printout would then show whether the particular employee is adequately trained and current and can be used in alerting employees of upcoming training which they need to attend.

Internal Communication and Employee Awareness

Closely related to the subject of training are the topics of internal communication and employee awareness. These topics are always important but have special significance in environmental management. This is due to the fact that environmental issues are often emotional topics to many employees. The right amount of information needs to be provided at the right time to the right employees. As you can imagine it is a challenge to get all these "rights" right. And then what is appropriate on one day may not be the next. If the proper communication does occur, employees will be aware and motivated to help protect the environment. Many things should be communicated to the employees, especially since they are responsible for protection of the environment.

Communication through Normal Management Channels

The traditional flow of information from one management level to the next is appropriate for certain environmental information. For example, managers should relay to their employees that they should come forward with waste minimization ideas. Managers should also instruct their employees to follow published procedures, for example, for handling empty chemical containers.

Communication through Internal Newsletters

The signing of a new environmental policy by the president would make a good story in the organization's newsletter. Another good news story would be when the organization phases out a major category of chemicals, such as Class I ozone-depleting substances. All employees should be aware of events of this magnitude and feel motivated and proud. In fact, a story of this magnitude should be considered for public release. This would build confidence in the company and raise employee morale.

Communication through Videos

Internally prepared videos are being used more frequently to relay environmental information to large groups of employees. If the video is introduced by the president or senior officer, the impact of the video can be maximized. Since the recommended length of most videos is no longer then 10 to 20 minutes, one would want to be careful that the information is essential. Videos are usually costly to prepare and can run from $10,000 to $50,000 for a professionally prepared 10-minute tape. Because of this high cost, dated material should be avoided or at least minimized.

Communication through Desk Drops, E-mail, Posters and Special Memos

Announcements concerning waste minimization programs, upcoming agency audits and other general interest information can be disseminated though desk drops, E-mail, posters and memos to large populations of employees. For this to be an effective method to communicate environmental information, it must be done frequently. It should be kept in mind that employees get busy producing the product or service and may not pay much attention to some of these types of communication.

Communication through Suggestion Boxes and Employee Hot Lines

Most of the communication methods mentioned above are from the environmental management department to the employees. In the case of suggestion boxes and employee hot lines, the communication is coming from the employees. Employees are a valuable source of information concerning actual impacts that are occurring in the field. They also will be the ones to come up with most of the realistic waste minimization ideas. Therefore this communication channel must be readily available for use by everyone. Rewards can be offered for suggestions.

Special Communications via Attorney

Certain environmental information should be relayed only to those with a need to know by an attorney. When communication is done in this way, the information is protected by the Attorney-Client Privilege doctrine. This means that if the whole situation ends up in court, the information would not have to be brought forward, which is important if it is supposition, guess work or theories.

Recognition

Positive and negative recognition must be parts of the overall plan. When employees perform waste minimization projects, for example, they should be given positive recognition. This not only rewards them for going beyond the call of duty but also helps encourage others to follow.

Negative recognition must also occur. Certain regulations, such as OSHA, require that an employee be given warning and even terminated if necessary if they violate the regulations. Two verbal warnings followed by two written warnings are usually given prior to termination, unless it is very serious and then termination can be immediate.

Performance appraisal forms should have a category for environmental protection efforts. This will allow direct recognition of the employee where it really counts, in their evaluation which affects salary. Depending on how well the employee has been working the performance appraisal form can show either negative or positive recognition in terms of environmental protection.

Changes/Continual Improvement

Organizational issues change rapidly in most organizations. It is therefore important to continually adjust the support systems in order to keep improving. Training and communication change especially quickly. As new information becomes available concerning environmental impacts, it must be used to upgrade training and should be considered in the communications that are sent to employees.

Chapter 8

Environmental Aspects, Effects and Impacts

Introduction

Without question, the most controversial part of the ISO standards deals with environmental effects/aspects/ impacts. Although there are differences, in this book effects have the same general meaning as impacts and aspects, that being measurable negative and positive changes to the air, water, land or natural resources.

BS 7750 specified that effects be identified and released to the public. The thought of assembling all effects into one document that could be released to the public struck terror in the hearts of many individuals, especially in the United States. Since that requirement has been removed, people have started to relax somewhat. Some still believe that assemblage and release of operational effects will be a certification requirement at some point in time.

Types to Be Analyzed

There are various types of effects that should be analyzed. The highest priority includes operational effects associated with the facility. The effects of the current operation should be compared with the previous operation. This will hopefully show statistically significant improvements and if not areas where work should occur. Operational effects need to be considered for start-up, normal and shutdown operations and emergencies.

The second priority of effects that should be assessed are those associated with the organization's product or service provided. The life-cycle analysis information presented in Chapter 18 gives suggestions concerning assessing a product's impact. Services can also affect the environment, such as a hazardous waste transporter or disposal site.

A third priority in terms of effects should be those associated with an organization's contractors. This is especially important if contractors provide many raw materials that require activities such as mining, smelting or manufacturing.

In other words, all types of significant effects related to the operation should be analyzed, especially those associated with the facility. Effects dealing with the product or service should also be considered; however, they do not carry the same importance in terms of ISO 14001 certification. Effects include those associated with planned actions, start-ups, normal operations (current), incidents, shutdowns and past operations.

A level of accuracy should be specified whenever the effects are being analyzed. The organization can, at this point, set its own level of accuracy. For example, the accuracy of predicting the effect of a planned action is much less than the accuracy associated with a current operation. This is due to the obvious reason that a real measurement can be made of a current operation.

Identification Procedure

The procedure that the organization uses to identify effects is a key requirement of the certification process. Because of this, the process should be clearly spelled out and appropriate for the organization in question. For example, the traditional risk assessment method of identifying and assessing effects might be very appropriate for certain situations and not for others. The procedure designed should allow identification of effects associated with all functions, activities and processes of the organization. A suggested generic effects identification procedure follows:

1. *Audits*—Routine audits should be performed by the environmental staff and possible effects/exposures noted. The managers and supervisors should also do audits and report any issues to the envi-

ronmental management staff. Special, nonroutine audits should also be done.

2. *Monitoring Data*—A review of normal monitoring data for high exposure levels should be carried out routinely. When out of the ordinary data is discovered, an action plan should be prepared to deal with the impact.

3. *Input from Others*—All employee complaints and suggestions, along with those from citizens and agencies, should be evaluated. It is possible that additional impacts may be identified in this way.

4. *Outside Audits*—Special audits should be done by outside consultants and, if concerns are identified, they should be noted as possible effects/exposures. The audits should be done by organizations that have no relationship or vested interest in the party being audited if the audit is to be truly independent. The audit should also be done by experts who know the industry, technology, local regulations, industry practice and regional laws.

Analysis Procedure

As mentioned above the normal risk assessment procedure may be appropriate for an organization to use to analyze certain identified effects or exposures. The risk assessment procedure for environmental effects includes a receptor characterization step, a hazard assessment step, an exposure assessment and a risk characterization step. If this fits the particular situation, it can be used. If, on the other hand, the traditional risk assessment procedure does not make sense, the organization has the flexibility to design their own effects-analysis procedure. Some common-sense procedural steps could include:

1. *Immediate Employee Impact*—Evaluation of the effect in terms of obvious and acute adverse impact on employee health and safety could be a first level of assessment.

2. *Immediate Environmental Impact*—Evaluation of the effect in terms of obvious and immediate adverse impact on the closest component of the environment, excluding employees, could be a second level of assessment.

3. *Regulatory Compliance*—Evaluation of the effect in terms of compliance with regulations might be the third level of assessment.

4. *Business Plan Alignment*—Evaluation of the effect in terms of compliance with organization standards and policies could be the fourth level of assessment. If the core business is not considered and addressed in the environmental control program, both may fail.

5. *Best Business Practice*—Evaluation of the effect in terms of whether it is controlled consistent with best industry practice might be the fifth level of assessment. Best business or industry practice commonly becomes law in many parts of the world.

6. *Long-Term Impacts*—Evaluation of the effect in terms of nonobvious and chronic impact on employees and the environment might be a sixth level of assessment. If it is possible that the chronic effect is significant, this level or phase of assessment would be conducted earlier in the process.

Prioritization

Once the above steps are completed, the effects should be prioritized so that the most significant ones can be addressed first. Most organizations do not have unlimited resources and therefore can't attack all effects at the same time. The effects that are causing obvious damage to the environment or human health should be tackled first. Effects or impacts that are costing the organization considerable money might be next on the priority list. Formulas for prioritizing waste streams to be minimized could be used to prioritize impacts.

Preparation of a Register

A Register of Effects is not expected to be required for certification at this time, however, it may be required in the near future. Even if it is not required, a proactive organization would already have the effects recorded and under control. In addition, it would be valuable for a register of effects-type document to be prepared and controlled by the individual at each operational site who has to minimize the effects as part of their everyday job. This key individual must have the major effects identified and roughly quantified in order to design and implement controls. At least an informal Register of Effects is needed for doing a meaningful job of environmental management.

Effects can be organized and presented in many different ways (see Figure 8–1 for one example). Effects can be considered from a total ecosystem or macro level or in terms of specific subcomponents of the ecosystem. A separate form could be used for each air, water and solid waste discharge or impact. This may result in a great number of forms, however, this would help keep impacts clearly identified and therefore easier to correct and track. The recommended subdivisions of the register of effects and a possible way to organize the sheets shown in Figure 8–1 are as follows:

1. *Air Emissions*—Point-source or end-of-stack discharges to the air are the first type of emission. This commonly includes contaminants such as particulates, sulfur oxides, nitrogen oxides, carbon oxides and other air pollutants. Fugitive discharges are usually in the form of particulates and volatile organic compounds and from large areas.

2. *Water Discharges*—Effluents which enter a sewer system, stream, ocean or lake from a discharge pipe are impacts that are regulated much more closely than nonpoint discharges. The impacts from nonpoint discharges are commonly from storm water runoff from broad areas and can be as significant since they are usually uncontrolled.

3. *Solid Nonhazardous and Hazardous Waste*—Solid waste is usually deposited in landfills. No matter how carefully the landfills are designed there may still be some environmental impact. Even the Class I hazardous waste landfill with its leachate collection systems, monitoring wells, liners and various other controls has been known to result in environmental impacts years after being authorized to operate.

4. *Other Impacts to Land*—In addition to the land impacts associated with solid waste mentioned in item 3, there are other significant impacts that should be considered. For example, impacts may be associated with use of the land for other than the preexisting use. Impacts to the land can also occur due to surface mining, deforestation and other activities.

REGISTER OF EFFECTS SUMMARY FORM

Date _____

Revision # _____

Prepared By _____

Approved By _____

Site Affected _____

Operation Causing Effect _____

Media Affected _____

Levels:

Uncontrolled	Controlled	Permitted
_____	_____	_____
_____	_____	_____
_____	_____	_____

Summary of Control _____

Status of Control (proposed, ongoing, etc.)_____

Cross-Reference to:

Regulation _____

Procedures Manual _____

Site Files _____

Audit Files_____

Other Files or Documents_____

Next Key Date When an Action is Required _____

Figure 8–1

5. *Use of Resources*—Use of water, fuels, energy and other nonrenewable resources are other possible impacts that may be important. Every organization, no matter how small, uses these types of resources. The amount of use should be quantified and minimized, even for the smallest organization.

6. *Aesthetic Effects*—Many social and health/safety impacts fit into the aesthetic category of impacts. The aesthetic effects may include noise, odor, dust, vibration and visual impacts.

The type of operation must be considered for each of these effects. For example, effects must be identified for normal operations, start-up operations, shutdown operations, incidents, planned activities and past activities. If an organization has several different sites, they may have operations that fit into several of these categories. The bottom line is that all phases of the operation must be considered.

A procedure to select the appropriate control to address the effect should be established. Whenever possible this procedure should specify best available technology (BAT). The control selected should also be cost effective, safe for the employees to operate and one that will obtain agency and public acceptance. Figure 8–2 presents a matrix or framework   that

CONTROL OPTION ANALYSIS MATRIX

Variables	Weight[a]	1	2	Options 3	4	5	6
Efficiency	5						
Cost	3						
Regulatory agency acceptance	4			Values[b]			
Public acceptance	2						
Safety to employee operators	4						

[a] Weight = Importance of the variable: 1 = Low Importance; 2 = High Importance. Score = Weight × Value.
[b] Degree of fit or match: 1 = Low; 10 = High

Figure 8–2

can be used to evaluate several possible control options. These options might be different treatment technologies. Most of Part III of this book presents controls which can be considered in order to minimize the effects which have been identified.

Presentation of Impacts

Impacts must be presented to various individuals in the organization in a way that is understandable. Senior management needs to know what impact the organization is having on the environment. They do not have time, however, to read through volumes of paperwork. One possible way of illustrating impacts in a way that can be understood quickly is shown in Figure 8–3. Note that the illustration could show all the impacts

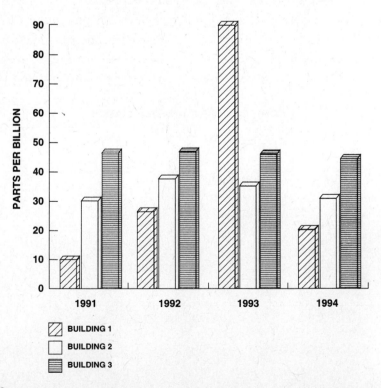

AVERAGE YEARLY AMOUNT OF MERCURY DISCHARGED FROM THREE BUILDINGS (EXAMPLE)

Figure 8–3

of the organization in summary form or details concerning one emission. For example, ppm/year of an air emission could be shown along with all the other impacts of the organization. The figure could present considerable detail such as the change in ppm of the air emission each month. There are hundreds of ways the data could be presented, so the challenge is to make it clear and meaningful.

Chapter 9
Requirements Imposed by Others/Regulations

Introduction

Environmental management may be one of the most heavily regulated disciplines in existence. Consequently, an organization can unwittingly violate a regulation resulting in fines, prison sentences, litigation, bad publicity, loss of certification and numerous other negative outcomes. To minimize such occurrences an organization must identify, analyze, record and comply with the regulations that apply to their operation. It is not uncommon for many industries in the United States to have to deal with 20 or 30 different environmental regulations at once. In many cases these regulations can be overlapping and conflicting.

Type of Requirements to Be Identified

Any regulation that applies to an organization's operation should be identified. These may include international, federal, state, regional and local regulations. At each level of government there will be several different regulations. In addition to the regulations, legislative acts and laws should also be identified.

In addition to the regulations there are various other requirements imposed upon an organization. These may include certification standards, corporate policies, customer agreements, court decisions, permits and other items. These requirements are also important and should be listed.

If all the enacted legislative and regulatory tracking is under control, then the organization may also want to list and track proposed legislation. This may include proposed bills, propositions, initiatives and regulations. A lot of what is proposed will never be enacted. Some of it will, however, and by tracking proposed legislation and providing comments, an organization may be able to help make it more meaningful and realistic. In addition, tracking proposed legislation gives the organization much more reaction time, sometimes as much as a year or two. This may make the difference between a cost-effective way to comply versus a costly, last minute method.

Types of regulatory requirements to be identified are at all government levels and apply to all segments of the environment. This includes legislation that covers the protection of air, water, land, wildlife, natural resources, employee health and safety, community awareness and other subject areas.

Procedure to Identify Applicable Regulatory Requirements

Considerable time can be wasted poring over volumes of regulations looking for those provisions that apply. The time spent can be minimized by concentrating on only the sections within the regulation that apply to the operation. The following procedure is suggested in order to comply with ISO requirements:

1. *Impacts*—Make a list of major areas of the environment impacted by the operation, that is, air, water, soil, wildlife and land. Also list major types of waste, such as hazardous waste, paper, aluminum, glass, that the operation generates.

2. *Agency Lists*—Check the numerous agency lists that are available. They are usually arranged by subjects similar to those in item 1. If your operation impacts that component of the environment or generates that type of waste, the corresponding agency probably has regulations that apply. If a commercially prepared agency list

is not available, check the phone book under federal, state, county and city governments. Within each of those major sections there will probably be specific agencies arranged by subject.

3. *Other Printed Sources*—Check a library or regulatory service for that type of regulation or the actual agency name and phone number. There are numerous documents which present the full regulation or a summary.

4. *Agency Contacts*—Contact the agency and get a copy of their regulations. Ask if they know of other regulations or agencies that might pertain to your operation. All levels of government, including international, federal, state, regional and local, should be checked to see if they have regulations that apply.

5. *Initial Document Check*—Check the index and preamble of the regulation and go to the relevant section. Bypass the background sections if the main body of the regulation does not apply.

6. *Detailed Document Check*—After speed reading the applicable section of the regulation, if it even remotely correlates to the operation, perform a more thorough analysis of the regulation.

Actual Record, Copy or Summary of the Regulation

According to the original recommendations found in BS 7750, a manual of regulations was to be prepared. At this point in time only the procedure for identification of regulations is needed. Even if ISO does not require a manual of regulations, it would still be a good idea for at least each organization site to keep a complete set of applicable regulations.

Corporate headquarters would not need complete sets of regulations from all sites, however, at least a one-page regulatory summary is recommended (see Figure 9–1). This summary sheet should clearly shows whether the regulation is international, federal, state, regional or local. Since most regulations are lengthy and complicated, the one-page summary sheet is valuable as a reminder and as an executive summary for upper management. These sheets could also be completed for proposed regulations as well, especially if they are likely to pass, and will have a big impact on the organization.

REGISTER OF REGULATIONS SUMMARY FORM

Date _____ Revision #_____

Prepared By_____ Approved By_____

Regulation Name _____

Regulation # and Section #_____

Administering Agency_____

Agency Address_____

Agency Contact and Phone Number _____

Implementation Date _____ Enforcement Date _____

General Subject Covered _____

Site(s) the Regulation Effects _____

Operation(s) Affected _____

Permit Required, Cost and Status _____

Plan Required, Cost and Status_____

Monitoring Required, Cost and Status _____

Equipment Required, Cost and Status _____

Building Modifications Required, Cost and Status _____

Other Important Requirements and Status _____

Location of the Regulation:

Regulation File at_____ Binder Located at_____

Computer File at_____

Cross-Reference to:

Register of Effects _____

Operations Procedure Manual _____

Environmental Management Manual_____

Figure 9–1

Appendix B shows one possible way to organize the regulation summary sheets or the actual regulations themselves. This would involve grouping all international regulations together, followed by federal, state and local. It should be noted whether the paper file, binder, or computer file contains the regulation and also whether it is physically located at corporate headquarters, the operating site or consultant location.

Chapter 10
Environmental Objectives and Targets

Introduction

A clear distinction should be made among policy, objectives and targets. A policy is a general, higher-level commitment and might include statements such as protection of the environment. An objective is a broad plan that will help achieve the policy, such as waste minimization. Targets are usually numerical goals that measure success of the objectives, such as 50 tons/year of waste recycled. Many organizations use these measurable goals as performance criteria.

The environmental objectives and targets should be consistent and incorporated into the organization's strategic plan. They should be compatible with the strategic plan of the organization as a whole or serious conflict will develop. Objectives and targets must all be consistent with each other and not conflict. The objectives and targets must also support legislative compliance, impact mitigation, business requirements and reasonable views of interested parties. Objectives and targets should be closely integrated with those of the rest of the organization. They definitely cannot be at cross-purposes or they will probably not be achieved.

Objectives and targets should be established together with timetables to achieve them (in phases if it is helpful to the project) and referenced to the preassessment report. The objectives and targets should take into consideration the active environmental projects, their status and their relationship to the new standards. New areas to be covered, as well as the need

83

for upgrading, downgrading or just continuing existing projects will also be indicated. All of this must be reviewed in relation to the new policy and/or revisions in the standards.

When the objectives are each clearly and precisely stated, a description of their proposed effect upon the environment, the organization, employees and the public is necessary. Planners would be wise to indicate those requirements of the international standards that are satisfied by each. Use of an independent, outside consultant may also be prudent at least during the planning phase.

Procedure to Specify and Communicate Objectives and Targets

 To achieve compliance with ISO each organization must have a way to specify meaningful objectives and targets. There is a lot of flexibility in the process used. One possibility follows:

1. *Subdivide the Policy Statement*—Break down the environmental policy statement into major subject components. Specify at least one concrete, measurable objective and several targets for each of the subjects identified.

2. *Designate Regulatory Requirements*—Identify major legislative requirements, such as obtaining a permit to initiate an operation. Make these objectives and specify a goal in terms of timing.

3. *Hazardous Waste Stream Impacts*—Set as an objective the reduction of major impacts of the operation associated with the significant hazardous waste streams. Make an annual percent reduction a target.

4. *Business Goals*—Select one of the organization's top goals that must be achieved to stay in operation, such as a certain profit margin. Determine how the environmental department can help support this goal, such as a target of 10 percent reduction in environmental expenses in the next year.

5. *General Comments Received*—Solicit and record views of interested parties. Also record and consider the unsolicited comments expressed by customers, agencies, environmental groups or other

individuals. Select the most reasonable views or suggestions and determine whether they are feasible. Also determine whether it will truly help the environment if they are implemented. If so, make the reasonable views into objectives and targets.

6. *Communication*—Communicate the objectives and targets to appropriate individuals in the organization.

7. *Continuous Improvements*—Modify or amend the objectives and targets when changes occur in the organization or to the product.

Mechanism for Achievement of the Objectives and Targets

ISO certification will require a description of the funding and procedures necessary to attain established objectives and targets. It is one thing to identify policies, objectives, targets and strategies, but achievement is what is really important. This is where creativity, energy and determination are especially important. A formal method for achievement is really secondary if the environmental manager has sufficient determination. One suggestion to help achieve the objectives and targets is to prepare a separate action plan for each. Figure 10–1 is an example of an action plan. Note that all the major steps or actions have been identified along with a person to perform the work. This designation of responsibility for achieving objectives and targets is essential. Also of critical importance are the completion or due dates for each step. When problems are encountered during execution of some of the steps, and there will be problems, the action plan must be adjusted if the target is going to be achieved.

For complicated objectives an action plan alone may not be enough. In these cases it is best to also utilize tools, such as project evaluation and review techniques (PERT). PERT systems/charts show all the interrelated tasks, along with early and late finishes, critical paths and other important information necessary to achieve the targets. All the horizontal lines in a PERT chart are individual tasks necessary to meet the target. The vertical lines illustrate interrelations among the tasks. The heavy line is the critical path, which must be completed on time or the entire target date will be

OBJECTIVE AND TARGET ACTION PLAN

Objective and Target(s) to Be Addressed:

_____

_____

_____

Date_____   Location_____

Responsible Person_____

	Specific Actions	By Whom	Proposed Completion Date	Actual Completion Date
1.				
2.				
3.				
4.				
5.				
6.				
7.				
8.				
9.				
10.				
11.				
12.				
13.				

Figure 10–1

missed. After input of certain data, some software systems will print out not only the PERT chart itself but also a status sheet for each subtask, along with early start, early finish, late start, late finish and other information.

Procedure for Dealing With Changes and Continuous Improvement

Objectives and targets should be adjusted when environmental and business conditions warrant. This will probably occur much more than a person would like and in many cases changes are unavoidable. There needs to be a balance between too many changes to targets and objectives and no changes. Either end of the spectrum can spell the end to meaningful and cost-effective environmental management.

In order to preserve the integrity of the overall environmental management program, the changes should be made from the microlevel and progress, if necessary, to the macrolevel. The first change should be to subtasks in the action plan for one objective. If this is not enough, then an adjustment should be made to several subtasks or an entire target. If all else fails, an entire objective may have to be upgraded or even replaced. When macrolevel changes are made to whole objectives or targets, it is very important to consider whether the change will impact other objectives and even the overall policy statement. This can happen in some situations due to the complex interrelationship of environmental issues.

Designation of Responsibility for Meeting Targets

Overall, it is important to have as much employee involvement in the decision making as possible. Flexibility in the achievement of targets and ability to make adjustments to the process involved are perfect examples of the responsibility the employees need. This requires that upper management trust and support all of their employees.

If responsibilities are not assigned, the targets will probably not be achieved. Therefore it is very important to designate key individuals at each organizational level who play a part in completion of the objective or target. For example, the president of the organization would be responsible for signing the environmental policy statement and supporting the concepts it promotes. The director or officer in charge of the environmental management department would be responsible for obtaining the human and financial resources necessary for complying with the policy statement and developing the objectives with input from as many employees as physically possible. An environmental engineer or specialist should be assigned responsibility for specific target(s).

Completion of most objectives and targets will also require assigning responsibility for completion of certain action steps to individuals outside of the environmental department. For example, individuals in the following departments are commonly assigned responsibility for helping to meet environmental objectives and targets:

- *Environmental Department*—Leadership or coordination of entire objective in terms of identification and tracking of action plan completion.

- *Engineering Department*—Implementation of many waste minimization projects. This is usually the case since the engineering department controls either the budget or the design effort.

- *Procurement/Purchasing Department*—Purchase of hazardous material services or products. The procurement department usually controls purchases or at least makes initial vendor contacts and agreement.

- *Legal Department*—Review of environmental laws and contracts. The legal department can help make some of the difficult interpretations that are usually encountered with most environmental laws, regulations and agreements.

- *Human Resources Department*—Assistance with environmental training and staffing requirements. In addition, the HR department may also need to be involved with corrective counseling if an employee is not following environmental control procedures or showing serious disregard for the environment.

■ *Public Affairs Department*—Liaison with media and customers. Environmental issues are a popular topic in the media, especially hazardous waste spills. Response to media and customer questions must be done carefully and with the advice and counsel of the public affairs department.

Examples of Environmental Objectives and Targets

There are literally thousands of possible environmental objectives and targets that an organization could pursue to help the environment. One example is shown in Figure 10–2. The real trick is picking truly meaningful and practical ones that support the organization's policy state-

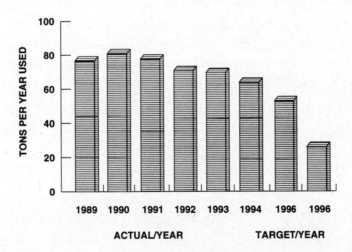

Figure 10–2

ment, legislation, impact mitigation and business requirements. Those that are appropriate and realistic for one organization would probably not be for other organizations. The following suggested objectives and targets are presented only to stir the creative imagination of the reader:

- Become ISO certified for environmental management by 1997.

- Reduce the purchase of x chemical by 50 percent within two years.

- Reduce the generation of y hazardous waste by 10 percent within one year.

- Correct all remaining open audit items within 60 days.

- Train x employees concerning hazardous waste within y days of employment.

- Build and/or update the chemical database by x date.

- Reduce discharge of x air pollutant by y ppm within 2 years.

- Reduce environmental expenditures by 10 percent by 1996.

Objective and Target Tracking System

Many objectives and targets may have minimum quantity or quality outcomes specified. If this is the case these criteria or standards must be identified and tracked so that it will be obvious when the objectives and targets have been met.

Part III
Environmental Management Programs and Operational Control Procedures

Once the elements presented in Part II are in place, an organization should next prepare and implement the actual environmental management programs and operational controls. These should apply, where appropriate, to products, services and activities associated with the organization. Examples of common programs, procedures and controls are presented in this part. It is important to emphasize, however, that these are examples and an organization should add, subtract or tailor what is presented here so that it is meaningful to their particular operation and is oriented toward their specific targets and objectives.

The programs, controls and procedures need to address the objectives and targets. In fact, these are the systems that will allow achievement of the objectives and targets. For this to really happen, the purpose and relationships to objectives and targets, responsibilities, time frames and resources should be specified in the systems. By relating or cross-referencing the systems to the objectives and targets, they will be meaningful and not just binders of paperwork sitting in bookcases. This will also encourage the responsible individual(s) to continuously improve the system so that the targets and objectives can be achieved in the most cost-effective manner possible.

As was the case with the initial elements presented in Part II, the environmental management programs, procedures and operational controls already in place in an organization should be used in the new ISO 14000 system. Missing systems needed for ISO 14001 certification could be added in with the current program if it is already well established and meaningful.

Whenever possible, best available technology (BAT) should be utilized in the operational controls. BAT may not be the most cost-effective technology for dealing with some problems and this must be considered as well. BAT is sometimes a regulatory requirement in some parts of the world for specific problems.

Chapter 11
Procurement and Vendor Controls

Introduction

Procurement and vendor controls of both environmental and nonenvironmental related purchases are very important in environmental management. Even the purchase of "normal" goods and services used to support the organization should be controlled. For example, suppliers should be asked whether their products can be recycled. Special care is needed when purchases involve chemicals or hazardous waste services. The suppliers should be encouraged or required to follow the same environmental management systems as the primary organization in question.

The environmental professional might be thinking that procurement or purchasing is the responsibility of another department and that they will take care of all the issues. Realistically, most procurement or purchasing departments are primarily concerned with only price, quality, delivery and minority status. Environmental concerns are usually not on the top of their list, or even on the list. It is therefore very important for the environmental professional(s) in the organization to work very closely with the procurement or purchasing department to make sure that environmental considerations presented in this chapter are addressed.

There are various ways to help ensure your contractors are doing their part to protect the environment. You could require that contractors that supply your chemicals, parts or services be ISO 14001 certified as well. Care must be taken, however, that this action does not create trade barriers or essentially eliminate most suppliers. Another way, but which is costly to your organization, is to audit all of them at frequent intervals. If this is not possible for financial reasons, then consider auditing only the vendors which have more of a potential to impact the environment.

Another way to learn about the vendor's concern for the environment, in addition to or in place of auditing them, is to review some of their environmental documentation. If the vendor is supplying a product, you can require an acceptable material safety data sheet (MSDS) or some other form of documentation concerning their product. This, however, will not give you any information about the environmental impacts of their manufacturing operation. To learn about the impacts of the vendor's operation, some of the following could be requested and reviewed. Most of the items of this list are available to the general public, but even if they are not, you can request them from your supplier and if they refuse, consider not doing business with them.

- Policy statement
- Hazardous material business plan
- SARA Title III chemical inventories
- Environmental permits
- Listing of environmental fines, cleanups, law suits

In addition to the normal reasons for procurement controls, the purchase of environmental services or hazardous materials adds considerable additional liability over other types of purchases. Environmental services, especially those dealing with hazardous waste investigation, remediation and disposal, usually require significant up-front proposal costs. Since all quoting vendors can't win the job, some may be bitter after spending considerable proposal money and claim unfair treatment. For these reasons it is very important to closely follow the procurement controls suggested in this chapter.

Treatment, storage and disposal facility (TSDF) vendors are a special kind of environmental contractor that should be managed with extra care. This is necessary since great liability is involved for both the organization generating hazardous waste and the TSDF that handles the waste. Hazard-

Table 11–1   Overview of the Procurement of Environmental Vendors

	Approximate Number of Vendors Potentially Involved
1. Request for qualifications	25
2. Qualifications received and reviewed	20
3. Qualified vendors	5
4. Send request for proposal or quote	4
5. Proposals/quotes received and reviewed	3
6. Best proposal/quote—vendor selected	1

ous waste that is handled properly and disposed of in a fully authorized Class I landfill in complete compliance with the regulations may later contaminate the environment. Unfortunately this happens more than the reader would like to know and usually results in significant cleanup costs for both the generator and the TSDF. Again, this underscores the attention that should be directed to the suggestions in this chapter.

An overview of the entire process is shown in Table 11–1. The numbers shown represent vendor companies. Note how the number of companies decreases from top to bottom. It may not be necessary or possible to get qualifications from up to 25 vendors as Table 11–1 illustrates. It is important to contact as many potentially qualified vendors as possible in order to get quotes from the best. In reality this may be as few as 5 or 10 vendors in some extremely technical fields.

Vendor Selection Process

1. Prequalification Phase

An organization should collect the credentials of as many vendors as physically possible. This would include their experience, financial stability, insurance, permits, resumes of key employees, environmental and quality programs and other information. The material is then screened and questioned, and a determination is made whether the vendor is qualified to perform the environmental service or supply the product containing a hazardous material. Screening criteria should be established so that all vendors are treated equally. Some suggested screening criteria are shown in Table 11–2. If these criteria are used for all vendors considered, then a fair comparison can be made. Probably the hardest of these criteria to quantify is the first one shown: experience. A shorter period of relevant experience is usually better than more years of general experience.

Table 11–2 Environmental Vendor Screening

Qualification Screening Criteria

Experience

Permits

Have an ISO 14001 certification

Financial stability/resources

Lack of fines and penalties

Personnel

References

Minority status

Proposal Screening Criteria

Depth or detail

Understanding illustrated

Creativity

Quality

Timeliness

Cost or quote

Responsiveness

This prequalification step applies to all environmentally related vendors. For example, it should be done for environmental consulting service vendors and hazardous material sale vendors. Prequalification should also be done for hazardous waste treatment, storage and disposal facilities (TSDFs). In the case of the latter, it is essential to visit the TSDF site and do an annual audit of the facility (see Appendix C). A check of the agency files in reference to the TSDF is also important. Once the qualification step has been completed for a particular vendor, they should be added to a tracking form which shows their approval status. See Figure 11–1 for a sample format. These forms should be provided to your organization's operating units with the requirement that they consider only approved vendors. It is important to keep reassessing the vendors and adjusting the form frequently, as their status can change quickly.

2. Request for Proposal or Quote

Only the vendors that meet the screening criteria described above and are approved should be invited to submit a proposal or quote. If the service or product is clearly understood by both the buyer and the seller, a quote is probably sufficient. If, however, the environmental service or product is complicated or not spelled out clearly, it is best to ask all qualified vendors for a proposal. The proposal should detail the product or steps the vendor will follow to complete the service.

The request for a proposal or quote should be made in writing and strict documentation controls followed. For example, all vendors should get identical requests that are mailed at the same time. If the vendors have questions, they should all receive the same answers in writing at the same time.

Usually vendors are given the opportunity to see the site or situation (walk-through). All vendors should visit the site at the same time so that everyone has the same opportunity and hears all questions and answers. The quote or proposal must then be submitted in writing and received in the specified time or it should be disqualified.

TREATMENT STORAGE
AND DISPOSAL FACILITY TRACKING FORMAT

Facility Information

Facility Name _____

Facility Address _____

Facility Contact Name
    and Phone Number _____

Service(s) Provided _____

EPA ID# _____

Sites Which Presently Use Vendor _____

Records Review

Date Records Were Last Checked _____

Permits Correct and Current _____

Any Notices of Violations, Clean-up Orders
    or Law Suits? _____

On-Site Inspections

Date of Last Site Inspection _____

Name of Inspector _____

Any Significant Problems Noted _____

Approval Status

Approved _____

Not Approved _____

Figure 11–1

3. Review of Vendor Submittal

Clear and measurable criteria for selecting the successful vendor must be in writing. There is a strong chance that this step might be questioned by an unsuccessful vendor. Some suggested criteria were shown in Table 11–2 for selecting a hazardous waste service vendor. The criteria should concentrate on the proposal, not the vendor's general qualifications. It is assumed at this point that all vendors invited to send a quote or proposal are already prequalified.

The major selection criteria for service proposals is usually cost; however, other factors are just as important. Far too often only the cost is considered and the job awarded to the low quoter. In some cases this can be a mistake as some vendors come in with a low quote and make up for it later with add-ons. The quality of the proposal is as important as cost, if not more so. The quality can be assessed by the level of understanding and depth shown in the proposal. Lack of errors is also a good measure of the quality that will be delivered during the actual job. The personnel proposed for the particular job are also of great importance. In some cases whether a job will be successful or not largely depends on the professionalism of the project manager assigned.

4. Selection of Vendor and Execution of Agreement

All vendors should be informed of the outcome and the successful vendor needs to sign the agreement covering the service or product. The agreement should have been part of the request for proposal or quote sent earlier so that last minute surprises or negotiations will not occur. Written agreements are an absolute requirement and can be either a one-job type agreement or a master-service type of agreement. Whichever type it is, liabilities must be clearly spelled out and fair to both the buyer and the vendor. Also certain of the buyer's environmental policies must be included in the agreement so that the vendor knows to follow them when they start work or shipping product.

Initiation of Work

Once all the agreement terms and conditions are worked out the environmental work can be started by the vendor or the product containing a hazardous material can be shipped. Right from the start the vendor must adhere to the buyer's agreement terms, especially compliance with environmental laws and policy. The organization's environmental professional(s) must closely monitor performance for a period of time since it is very hard for a vendor to remember all the proposal and agreement terms and conditions.

A good ongoing relationship is as important with environmental vendors as it is with all other types of vendors. You don't want an environmental vendor cutting corners because they underquoted or because they did not understand the original scope of the job. Also, if a big liability problem develops, no matter what the agreement states, both the buyer and the vendor are going to suffer. A team effort or partnership with the vendor during the project will reduce the amount of suffering for both parties if a problem develops during delivery.

Procedure for Dealing with Existing Product Suppliers

The preceding pages in this chapter dealt primarily with procurement of goods and services from new suppliers. In this section the emphasis will be more on the procurement of environmentally friendly goods from current suppliers. Buyers these days are demanding more than just low price and high quality. Many also want "green" products or the ones that do not have an impact on the environment. The following procedure will help achieve this environmental goal.

1. Request Product Composition Data

If the manufacturer has not already sent a material safety data sheet (MSDS), ask them to do so. Other types of data would also be acceptable as long as the chemicals are identified along with the percent composition. For example, you should know whether the product you are buying contains fire retardants such as asbestos and polybrominated biphenyls (PBBs). You do not want to buy a product containing a banned or soon to be banned chemical as is the case with many of the fire retardants.

2. Request Data Concerning Chemicals Used in the Manufacturing Process

Ask your suppliers whether their products require chemicals during the manufacturing process. The answer is usually yes, numerous chemicals were used. This is not necessarily a bad answer, but depends upon the chemicals used. Therefore the question may have to be qualified somewhat or chemicals of concern listed. If many suppliers are involved it might be easier to send a questionnaire with the chemicals of concern identified. For example the list might include ozone depleting substances (ODS), especially Class I ODS. World-wide attention is being focused on ODS and some customers are requiring that none be used in their suppliers' manufacturing process.

3. Request Information Concerning Product Life Cycle

Everything mentioned in the section entitled "Design For Environment/Life-Cycle Analysis" in Chapter 18 could apply to the parts supplied by the organization's vendors. For example, the buyer could ask about their supplier's raw materials and the impact on the environment when they were produced. The impacts of the supplier's part while it is incorporated in the main product, such as emissions, energy and water utilization are also a consideration as is the disposition of the part at the end of the product's life. Can it be reused or recycled or must it take up valuable landfill space? If so does it contain some chemicals, such as heavy metals, which might leach into the environment after it lays in the landfill for a number of years? All these issues could be a consideration in the decision concerning continued purchase of the part.

4. Contractual Modifications

The buyer of a product should try to renegotiate in the purchase agreement as many reasonable environmental safeguards as possible. Some buyers go overboard, however, and demand excessive and impractical environmental conditions, which prevents signing or are ignored by the seller who will sign anything in order to sell the product.

Changes/Continual Improvement

Vendor interactions should be improved and upgraded on an ongoing basis. Even if a good business relationship exists at one point in time, it is possible that problems may develop because of changes in personnel and business conditions. When these changes occur, the procedures discussed throughout the chapter will need to be implemented or adjusted.

Chapter 12

Process, Equipment and Chemical Approval and Tracking

Introduction

Many processes require chemicals, water, electricity and gas or create discharges which end up in the sewer, air, landfill or a water body. Because of this it is important that the environmental department does a review before implementation and provides recommendations. It may be necessary to insist that some of the important recommendations be accepted before signing off on the proposed plans. If drawings, plans and designs are reviewed prior to implementation, many environmental problems can be eliminated before they occur. This review process should apply to new and modified process and equipment plans.

Process, Equipment and Chemical Review and Approval Procedure

An overview of the entire process in shown in Figure 12–1. The key part is in the center of the figure and is entitled "Environmental and Safety Approval." The drawings or plans should be reviewed to ensure compliance with the policy statement, targets, objectives and regulations. If a potential problem is noted before the chemicals are introduced into a process or equipment is installed, it can be corrected in a more cost-effective way than if it is done after the fact. Funds, time and labor can be conserved by correcting discrepancies before the process is installed or equipment purchased. Suggested review and approval steps include:

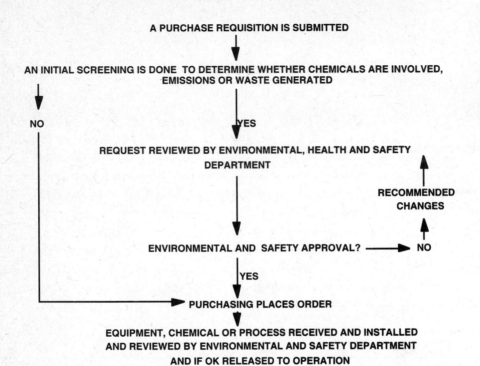

NEW PROCESS, CHEMICAL AND EQUIPMENT INTRODUCTION FLOWCHART

A PURCHASE REQUISITION IS SUBMITTED

AN INITIAL SCREENING IS DONE TO DETERMINE WHETHER CHEMICALS ARE INVOLVED, EMISSIONS OR WASTE GENERATED

NO　　　　**YES**

REQUEST REVIEWED BY ENVIRONMENTAL, HEALTH AND SAFETY DEPARTMENT

RECOMMENDED CHANGES

ENVIRONMENTAL AND SAFETY APPROVAL? ⟶ **NO**

YES

PURCHASING PLACES ORDER

EQUIPMENT, CHEMICAL OR PROCESS RECEIVED AND INSTALLED AND REVIEWED BY ENVIRONMENTAL AND SAFETY DEPARTMENT AND IF OK RELEASED TO OPERATION

Figure 12–1

1. Advance Memo

Send memos in advance to all departments that might implement a new chemical, process or install a new piece of equipment. Ask the department to send a copy of any drawing or design document to your office for review as soon as a draft is available. See Figure 12–2 for an example. This particular memo should be sent several times as a reminder.

2. Narrowing of Request

The environmental manager should not be surprised if all drawings are not sent. Even 50 percent is an optimistic return in some organizations.

ADVANCE MEMO (EXAMPLE)

Date:

To:

From:

Subject: Review of Drawings, Plans and Designs

All drawings, plans and designs for new or modified systems must be sent to the Environmental, Health and Safety Department for review and approval. This will ensure that environmental, health and safety issues are appropriately addressed in a cost-effective manner. Drawings, plans and designs that pertain to the following are of special importance and should be sent quickly and in complete detail:

1. Processes that will use chemicals

2. Processes that will generate emissions to the air, water or land

3. Processes that will generate waste, both hazardous and nonhazardous

4. Processes that will utilize energy, water or other natural resources

If there is any question concerning the above, the information should be sent, since most processes fit one of the four categories. Thank you in advance for your support and if you have questions please call_____.

Figure 12–2

If resistance is encountered, qualify the request in more detail by asking for designs that meet the following considerations (roughly prioritized):

■ *Uses a Chemical in the Process*—The most important consideration in terms of design review are plans that involve chemicals. These types of plans, no matter what the chemical, should be reviewed by the environmental, health and safety professional(s).

■ *Discharges to the Air, Water or Land*—It may not be obvious during the design phase whether there will be discharges to the air, water or land. If it occurs to anyone that there may be discharges, however, the designs should have an environmental review. The point-source or end-of-pipe/stack discharges are easier to predict and these plans should be sent to the environmental department for sure. Nonpoint-source discharges, such as release of volatile organic compounds from a large area, are harder to identify but should also be reviewed.

■ *Generates a Hazardous Waste*—The design department may not know whether a waste product is hazardous or not. If there is any question, the waste should be considered hazardous and the plans reviewed by the environmental, health and safety professional(s).

■ *Uses a Chemical in the Cleaning or Maintenance of the Process or Equipment*—Maintenance chemicals usually cause less impact; however, if the design calls for them, the plans should have an environmental review. Some maintenance or cleaning chemicals, such as ozone depleting substances (ODS) used in cleaning parts, do result in impacts and must be reviewed before the process starts.

■ *Generates a Solid Waste Such as Paper, Aluminum, Glass or Plastic*—With the growing shortage of sanitary landfills there will soon be few options for the handling of solid waste other than reuse and recycling. Increased disposal costs necessitate that the environmental person review new plans with this in mind.

■ *Utilizes Considerable Energy, Water or Other Natural Resource*—In many parts of the world some natural resources are in extremely short supply. All of these resources should be conserved since most are nonrenewable. Therefore they should be considered during the design phase and every attempt made to minimize their use. This is important in terms of the environment and operating costs.

3. Review of Materials

When the drawing or design arrives, it should be reviewed immediately to assure compliance with environmental, health and safety requirements. If not enough information has been sent, transmit a request for additional information (see Figure 12–3). The review must be done promptly so that the design process is not held up. If delays become commonplace because of the environmental, health and safety review, plans may not be sent over in the future.

4. Field Review

If anything of concern is found in the paperwork, a personal visit to the design engineer and location should be made to analyze the situation in greater detail. The field visit should be made if at all possible regardless of what the paperwork review uncovered. It will be helpful if the reviewer can actually see the operation or a similar operation where the proposed process or equipment is to fit.

5. Recommendations

Suggestions concerning ways to minimize the proposed use of the chemicals, natural resources, emissions or the impacts should be made in writing. Waste minimization concepts, engineering and administrative controls and personnel protective equipment are all possible approaches that could be recommended. It is very important to do more than just identify potential problems. Work with the design engineers and anyone else who can help and come up with a few possible solutions as well.

6. Sign-off

A sign-off or recommendations should be made concerning the proposed chemical, process or equipment by the environmental, health and

REQUEST FOR ADDITIONAL INFORMATION CONCERNING NEW PROCESSES, EQUIPMENT OR CHEMICALS

Date:

To:

From: Environmental, Health and Safety Department

Subject: Request For Additional Information

Please provide the following information concerning the proposed system:

System Name _____ System Location_____
Project Manager _____ Phone Number _____

Chemical Usage Information
(Attach an MSDS from the manufacturer)

Chemicals Which Will Be Involved _____
Chemical Percentages and Volumes_____
Manufacturer _____
Estimated Usage _____

Waste Generation

Type of Waste Generated _____
Volume of Waste Generated_____
Proposed Method of Treatment _____
Proposed Method of Disposal _____

Energy Utilization

Amount of Planned Water Usage _____
Amount of Planned Electricity or Gas Usage _____

Figure 12–3

safety department (see Figure 12–4). This should be done in writing within a specified and reasonable time period. When it is not done in writing according to a schedule, it is likely that the environmental, health and safety input will not be incorporated.

7. Ordering

When it is time to order the chemical or equipment it is necessary to work closely with the purchasing department. This will allow one additional layer of screening and control to ensure that only authorized vendors are used. In most cases the environmental manager starts to lose control at this point since the chemical or equipment is usually ordered by someone else in the organization.

8. Follow-up

It is essential for the environmental manager to follow up. Double checking should be scheduled to ensure the recommendations were followed and that the overall design for the process or equipment has not changed radically. This includes ongoing discussions with the design engineer, purchasing department, receiving department and with the individuals doing the actual installation or using the chemicals.

Procedure for Tracking Chemicals

 Tracking chemicals is an important component of most environmental management systems. Being aware of chemicals which are planned, purchased, stored and used and which become hazardous wastes is the essence of the system. Key components of the tracking system include an inventory along with important information about the chemical(s) and a mass balance.

If some of the same chemicals are used at different locations for the organization, it might be more cost-effective to have a central or corporatewide software system. Site-specific information would have to be easily entered, however, from field locations. A general procedure for chemical tracking follows.

DRAWING, PROCESS OR PLAN SIGN-OFF FORM

Drawing or Plan Number _____

Name of Proposed System _____

Project Manager _____

DATE_____   LOCATION_____

_____

Reviewed and Approved By (If stipulations or conditions are made, they will be attached):

_____   _____
Environmental Department   Date

_____   _____
Safety Department   Date

_____   _____
Facilities Department   Date

_____   _____
Quality Department   Date

_____   _____
Engineering Department   Date

_____   _____
Production Department   Date

Figure 12–4

1. Identify Responsible Individual

Designate a person who is responsible for all chemical tracking at the site. Many people may be involved; however, one individual should coordinate the overall effort for the entire site.

2. Inventory

Inventory the chemicals and enter this information into a software system. For example, it should be indicated whether the chemical is planned or on-site. The quantity of the chemical in use, storage and being discharged needs to be recorded on an ongoing basis. The discharged amount would be in accordance with permits and regulations and include going to sewers, water bodies, air and land fill. The amounts presently being recycled, reused or sold should also be added into the inventory and so designated. Many of these figures could be obtained from manifests, SARA Form R reports and other sources.

3. Other Information

Additional information should be entered into the software system; however, this may require some research into material safety data sheets or calls to the manufacturer. For example, the composition of the chemical, Chemical Abstract System (CAS) number and hazard rating information should also be added.

4. Ongoing Review and Entry of Information

There should be ongoing review of invoices, shipping documents, manifests, plans and other information, and this information should be added to the software system. The system should track, balance and account for all chemicals. This helps to ensure that some have not leaked or been spilled from their storage tanks, for example.

5. Enforcement of Chemical Purchase Procedures

As all the numbers are obtained, it may be found that some individuals are not following the procedures. When this happens, they must be reminded of the procedure, and, if necessary, disciplined.

6. When the Numbers Don't Add Up

If the mass balance shows that there are significant volumes of chemicals unaccounted for, some additional research will be needed. If it is not just an error in calculation, it might be a leaking storage tank or a spill.

Procedure for Handling Empty Chemical Containers

Since most empty chemical containers may still have some residue, they must be tracked and handled properly to minimize injury and impact to the environment. In many cases, the residues in the containers would have a hazardous waste classification. Organizations should have procedures for the tracking and handling of empty chemical containers, whether they are common chemicals, such as photocopy toners, or the highly toxic chemicals.

1. Training

Employees should be trained in the handling of empty chemical containers. For example, the employees must know that they are to have PPE when the procedure is carried out. If the employees don't understand or appreciate what they are doing, there will be containers going out in the trash with residual chemicals. This may result in injury, environmental damage and heavy fines.

2. Removing of Residual/Residue

Remove as much of the chemical as possible. This may require inversion for a period of time, scraping or chipping. Drain racks are commonly used for this phase of the process.

3. Handling of Residue

Use the residue in the plant operation if possible. If this cannot occur then the residue will probably have to be treated as hazardous waste and should not be placed in the common sanitary landfill. If liners are present they are also usually treated as hazardous waste.

4. Washing of Container

Depending on the residue and whether the company has washing and treatment equipment, it may be possible to wash out the container and cap. This washing is usually done three times (triple rinse). The wash solution may have to be treated as hazardous waste. In some situations, this washing process may require a treatment permit.

5. Testing of Container

The container should be tested after the washing. For example, if an acid or base was previously in the container, pH paper should be used. pH meters are another option to ensure adequate washing has occurred.

6. Inversion of Container

After washing the container, it should be inverted. A drain rack will allow all residual fluid to exit the container.

7. Return of Container

If possible, the container should then be returned to the chemical supplier, broker or recycler. This is preferable to land disposal in terms of the environment and liabilities.

Chapter 13
Emergency Procedures

Introduction

No matter how many controls are implemented, it is impossible to completely eliminate every problem or emergency. These could include accidents, spills, injuries and disasters. In order to prepare for these unforeseen events, detailed emergency procedures should be written before any emergency occurs. Plans that are prepared to address these situations have many names; however, the two most common are the emergency response plan and the disaster recovery plan. In an oversimplification, emergency response plans deal with providing immediate help to individuals and the environment whereas disaster recovery plans specify how to get the business back up and running.

Emergency Response Plan

An emergency response plan is usually designed to cover immediate protection of employees and the environment. It concentrates on the actions to be taken in the first few hours of a crisis. For example, the immediate evacuation of personnel and cleanup of a spill or control of a fire are common components. Execution of the plan is usually under the direction of the emergency response team (ERT).

Preparation and Distribution

The emergency plan should be prepared by the site environmental, health and safety individuals who have knowledge of site conditions and regulations. State and local agencies have specific requirements that should be incorporated. Security, facilities, legal, and human resource departments should also contribute to the plan or at least complete a review. The emergency response team itself must also be involved in the preparation of the plan or at least in the continual upgrade of the plan over time so that they are intimately familiar with it and feel like contributing authors.

The plan should be amended when important components become outdated or business and regulatory changes dictate. Whenever changes are made, a revision date and number must be noted. In addition to the revision date each plan/binder should be numbered for tracking purposes. All changed pages should be sent to all plan holders with a controlled copy. A controlled copy is a numbered copy and the holder is a person who automatically gets revisions. When a plan holder receives an update, they must verify receipt, such as by signing and returning a transfer memo to the department coordinating the plan.

Any plan issued without a number would be considered an uncontrolled copy. It is the responsibility of the holder of any uncontrolled copy to request updates. Any copies made of uncontrolled plans or portions of plans should be clearly marked "Uncontrolled Copy."

Many different people in the organization should be given a copy of the plan and everyone should have easy access, if not their own copy. At least one copy should be in every building; it is usually located at the reception desk, guard station or in a wall-mounted box near the front door. In addition, the following individuals should have their own "controlled copy":

- Each ERT member
- Safety committee chair
- Site environmental, health and safety representative(s)
- Local fire department
- Local hospital
- Corporate environmental, health and safety

Up-Front Activities and Components which Should Be Prepared before an Emergency

All emergency response plans are unique to a particular operation. They must be specific if they are going to be useful in an emergency. There are a few key up-front elements in most emergency response plans and they are as follows:

1. Emergency Response Team (ERT)

The ERT should be composed of employees who have knowledge or can be trained in responding to emergencies such as fires, explosions, chemical spills, and so forth. Designate an adequate number of employees to serve as emergency response team (ERT) members. There must be teams for all shifts and team members need to have alternates. Each team needs a team leader.

Most organizations will require each functional department to designate one person to serve on the ERT. If this does not generate an adequate number of ERTs, then they could be assigned by building. Due to the great amount of training required, it is most cost-effective to require the ERT member to serve at least two years and longer if they desire.

One key member of the ERT is the incident commander or team leader. This person must be selected very carefully since they may have to make critical decisions in high-pressure, crisis situations. Some of the decisions may have significant impacts on employee(s), the environment and the business. The person selected should be a clear thinker, calm, educated and have the ability to lead. In other words, the ideal employee.

In the "lean and mean" organizations that are so common in industry today it is sometimes hard to get an adequate number of ERT members. Every department seems to be understaffed and finds it difficult to designate a representative to join the ERT. To deal with this problem the environmental, health and safety department should first put out a request for volunteers to join the ERT. For obvious reasons the individuals who want to be on the ERT are much more valuable than the ones who are assigned. If enough volunteers are not obtained, then the manager of each department must designate who will be on the ERT. It may be necessary for the environmental, health and safety department to send a copy of the policy statement or some other document that specifies the requirement for an ERT that has been signed by upper management.

2. Personnel Protective Equipment

Personnel protective equipment (PPE) should be stationed in strategic locations for the ERT. Depending on the chemicals in the facility, it may include self-contained breathing apparatus (SCBA), chemical suits, gloves, boots, respirators and cartridges and duct tape. Depending on the PPE used, it may be necessary to do fit tests, such as for respirators, before an emergency occurs.

3. Chemical Spill Cleanup Supplies

Before an emergency occurs it is necessary to purchase cleanup supplies for spills and station the supplies in high-risk areas. The type of chemical dictates the supplies needed; however, these commonly include absorbent pillows, pads or booms, acid neutralizers, base neutralizers, pH paper, disposal bags or drums, hazardous waste labels, barricade tape, storm drain covers, tool kits, brooms, shovels and squeegees.

4. Training

The ERT members must be trained in how to handle various situations such as spills, fires, injuries, earthquakes, and extreme weather problems. First aid and CPR training are the first instructions the ERTs should get. This should be followed by hazardous waste operations and emergency response (HAZWOPER) first responder training which is required by the Occupational Safety and Health Administration (OSHA) for all ERTs. Depending on the chemicals in the facility the training must last from 24 to 40 hours. Subjects covered include respiratory protection, toxicology, incident command systems, spill cleanup procedures, handling of drum emergencies, hazard classification, MSDS understanding, hazard identification and assessment, PPE, monitoring equipment, first aid, fire control, Department of Transportation (DOT) Emergency Response Guide, decontamination and numerous other general and site-specific topics.

It is very important that management support the ERT training. The supervisors must allow time for training and insist that their employees are properly trained in the ERT functions. Without this type of strong support the ERT may not be able to find the time for the training. This could result in injury to ERT members and possibly other employees. The site environmental, health and safety representative(s) and ERT incident commander should also encourage and document that the necessary training is occurring.

Special training must be given to the ERT incident commander or leader. This person will be making many critical decisions, some of which may have significant impact on health, safety, environment and the business. Therefore this key individual should attend special classes, in addition to the ones listed above, which can be found at some universities and other locations. Incident commander class, chemistry, and advanced HAZWOPER are examples of specialized training that should be completed by the leader.

5. ERT Drills

The ERT must go through drills to practice the skills they learned during training. For example, they should have mock spills to ensure they are following the right procedures. The drills should occur at least every other month, with a discussion of the success and problems encountered. The drill could occur during a scheduled monthly meeting of the ERT and occasionally be unannounced.

6. Physicals

All members of the ERT must undergo physicals and respirator fit tests. Medical examinations must be required to minimize the chance that an ERT member cannot perform the function because of some physical limitation. A doctor should certify that each ERT member is physically fit to participate in ERT activities.

7. ERT Communication

Pagers or some other form of emergency communication equipment must be provided to the ERT. The ERT members must each have a pager, radio or cellular phone so that they can be summoned to the scene quickly. Their pager numbers need to be given to at least the security command post, reception desks, internal operators, site environmental/ health and safety representative(s). It is also a good idea to provide some type of emergency communication device to the lead site environmental/ health/safety representative, security guard and site nurse since they are valuable resources to the ERT during an emergency.

8. Emergency Response Plan

The entire emergency response plan needs to be prepared before an emergency occurs. This entire section of Chapter 13 could be considered a generic emergency response plan. All the reader would need to do is insert site-specific information. The plan should be updated at least once a year or sooner if significant changes have occurred.

9. Availability of the Team

At a minimum the ERT must be available during operating hours of the facility. For continuous operation this would mean that the ERT should be on-site 24 hours a day, 7 days a week. Obviously there would need to be several teams to cover all the shifts. Even if the facility does not operate continuously, the ERT should be on call in case a leaking tank or other problem occurs during off-hours. Security guards commonly help out with some ERT functions for off-shift operations if no one else is around. If this is the case, they should be familiar with the above mentioned steps.

10. Designated Internal Emergency Phone Number

A dedicated site emergency phone number must be set up that can be used from any in-house telephone. It is best if the phone number is simple to remember, such as 2222. When someone calls, they need to be immediately connected to the security command post that will dispatch the ERT and/or proper community agency (fire, police ambulance, etc.). Callers should be instructed to stay on the phone long enough, if it is safe to do so, to answer applicable questions such as those shown in Figure 13–1. If the caller is physically or psychologically unable to answer many questions, then the security command post should concentrate on at least getting the caller's location and concern. While this is being relayed to the ERT, more information can be asked of the caller.

11. External Emergency Phone Numbers

Phone numbers and guidelines must be provided concerning calls to the police, fire department and ambulance. The guidelines are very important since there are many gray-area situations when the security command will be unsure whether it is really an emergency or not. If there is any question it should be assumed it is an emergency and the appropriate service called. For example, the following would be called immediately by the security command post for life threatening conditions:

- 911 or ambulance

- Fire department

All other calls would be made by or under the direction of the ERT Incident Coordinator/Team Leader. If possible, these calls would be made after a brief discussion with the site environmental, health and safety representative(s) to determine applicability and seriousness (not arranged in order of priority):

- *State Office of Emergency Services*—Most emergencies

- *Wastewater Treatment Plant*—Spills into the sewer

- *Local and State Water Quality Office*—Spills into streams, lakes, drainages

- *Local and State Environmental Office*—Spills over the reportable quantity

- *Fish and Game Office*—Spills that may impact wildlife

CALL RECORD SHEET EMERGENCY INFORMATION

Name of Caller? _____

Date of Call?_____ Time of Call?_____

Location of Emergency?_____ Phone # of Caller? _____

Medical Emergencies

Description of Problem? _____

Suspected Heart Attack?_____ Loss of Consciousness?_____

Broken Bones?_____ Bleeding?_____ Seizures?_____

Breathing Problems?_____ Approximate Age of Victim?_____

Fire Emergencies

Flames Present?_____ Smoke?_____ People Injured?_____

Approximate Area Involved? _____

Suspected Source? _____

Has the Area Been Evacuated? _____

Bomb Threat Emergencies

Where Is the Bomb Located?_____

What Time Will it Go Off? _____

Area Been Evacuated?_____ Type of Bomb? _____

Chemical Spill

Name of Chemical?_____ Volume Spilled?_____

Injuries?_____ Area Been Evacuated? _____

What Type of Hazard Does the Chemical Pose? _____

Natural Disasters/Earthquakes

Type of Disaster? _____

Extent of Damage? _____

Injuries? _____

Any Assistance On-Site? _____

Figure 13–1

■ *Air Pollution Office*—Releases or spills that become airborne

■ *Local Gas and Electric Company*—Gas leaks and power failures

■ *Highway Patrol*—Emergencies on roadways (spills, etc.)

■ *Police Department*—Local emergencies

■ *Hospital*—Medical emergencies

■ *Local Clinic or On-Site Nurse*—Medical emergencies

■ *Fire Department*—Fires, explosions, spills

■ *Local Hazardous Waste Cleanup Service*—Spills

■ *Chemtrec (800-424-9300)*—Chemical spills

■ *Company Public Relations Office*—Any emergency

■ *Corporate Offices of Environmental, Health, Safety and Security* — Any emergency

12. Evacuation Maps

Up-to-date evacuation maps must be prepared and posted in numerous locations in the facilities. At a minimum, these maps must show the closest exit, backup exits, and assembly points. It is advisable that they also show the location of the ER plan, security/reception desk, fire extinguishers, eye wash, emergency showers, spill supplies, first aid supplies and other key elements. Employees should be told to memorize the primary route identified for them and a backup route in case the primary exit is inaccessible.

13. Public Address System

Some type of communication system needs to be ready in case of an emergency. Emergency alarms with battery backup that can be heard everywhere in the facility should be required in 99 percent of all settings. In the 1 percent of the buildings where this is unrealistic because of extremely small size or some other reason, there should be some other reliable form of emergency communication. For example, hand-held megaphones or some other public address system could be used if they comply with local regulations. Whatever system is selected it must be heard in every area of the facility that employees might frequent. This includes remote areas, bathrooms, break areas and noisy areas. The emergency communication system should be tested monthly to verify that it is operating properly.

14. Outside Assembly Points

Several predesignated outside assembly points should be marked and employees instructed where to assemble in the event of an emergency. The supervisors should be told that this is where they take a roll call once the evacuation has occurred. To do this efficiently the supervisor must be able to quickly tell who is on which shift and on vacation and sick leave.

15. Other Emergency Equipment

In addition to the spill cleanup supplies, radios and PPE already mentioned, which are the most important, there is other emergency equipment that should be obtained as well. Safety showers, eye wash stations, fire extinguishers, first aid supplies, bloodborne pathogen kits, stretchers, backboards, oxygen and decontamination equipment are examples of other useful equipment.

16. Practice Evacuations

At least once a year the entire employee population should have an evacuation drill. If one facilitywide drill is too disruptive to production, then each department could do a separate drill. Also, if a few employees have to remain behind to maintain a critical process, they should be walked through their own drill after their shift. If at all possible, it is best to have the entire facility participate together in one drill, as would be the case in a real emergency.

During the drill and real emergencies, handicapped employees must be assisted from the area. The supervisor should assign two employees to assist each handicapped individual during the evacuation. This designation should be made well in advance of drills and emergencies. The individuals assigned must be physically able to carry out the handicapped employees, if needed.

Based upon notes taken by the ERT during the drill, a report should be given to all supervisors concerning success of the drill. Suggestions should be made to decrease evacuation time and increase overall control. Any special problems encountered should be noted along with corrective actions. Once the supervisors get the report, they should go over it with all their employees.

Activities During an Emergency

The ER team should be prepared and able to respond quickly, efficiently and safely to an almost infinite number of possible situations that could occur in an actual emergency. It would be impossible to list all recommended actions since what is appropriate in one situation may not be advisable in another. The following are therefore only examples. An attempt was made to present these suggestions roughly in the order in which they commonly occur. In certain situations, however, it may be necessary to reorder the following steps or do some in parallel.

1. Notification

The ERT is notified of an emergency by the security command center or some other source and the ERT assembles near the site of the emergency in a safe area. The ERT notification may be made by individual pagers, radios or a public address system. The public address system should be a last resort as this may cause employees to panic and at a time when the ERT would not yet be assembled to help manage the panic situation.

2. Evacuation

The ERT would sound the alarm and clear employees out of the affected area if there were an immediate threat to human life. The decision to evacuate employees should be made by the ERT incident commander, with input from as many knowledgeable individuals, such as the area supervisor, as time will allow. Employees should be told to move out of the area in an orderly manner through designated routes identified on the evacuation maps.

Once the alarm or announcement has been made, the ERT and supervisors should ensure that the evacuation is proceeding smoothly. For example, employees should not be panicking, not using elevators and not picking up personal belongings. As the supervisor exits the area with their employees, they should make a quick check of rest rooms or other areas where an employee might still remain.

3. Accounting for Employees at the Assembly Points

It is the supervisors' responsibility to account for all their employees at the assembly point, including those out on medical leave and vacation. If any employees are missing, the ERT incident commander must be notified immediately of the name and last known location. Only a team of two trained ERTs wearing proper PPE can reenter to search for a missing employee. Employees must be told not to try to reenter the area until the all-clear signal is given by the ERT incident commander.

4. Assessment of the Emergency

After a brief interview with the involved employees, the ERT will put on PPE and inspect the area to make sure all employees are out and make an assessment of the emergency situation. The buddy system must be used during this assessment. The substance spilled or other cause of the emergency should be clearly identified by looking at labels, using hand-held meters or other methods.

5. Remove Injured Employees

If injured employees are found, they should be carefully moved out of the area of concern only by the ERT who must be wearing appropriate PPE. Depending on the injury it may be necessary to wait until an ambulance arrives with the proper equipment to move the injured person(s).

6. Initial Calls to Outside Resources and Agencies

If immediate assistance is needed, the ERT incident commander will instruct whom to call from the preceding list. This may include the fire department, ambulance and other emergency type agencies. The fire department is commonly called, just as a contingency, even if the ERT seems to have the incident under control.

7. Shutdown of Certain Utilities and Services

During an emergency it may be necessary to shut down gas, electricity, water or other services. The incident commander of the ERT will make this decision with input from others, such as the facilities department. Care must be taken to not shut down too much as this may hamper resolution of the emergency and cause serious disruption to the business.

8. Erect Barricades

Barricades or barriers establish a zone of isolation that should prohibit entry by everyone except the ERT. When creating the zone of isolation the ERT should consider that some vapors may travel great distances very fast. If this is the case a considerable distance downwind may have to be evacuated and barriers set up. The police and other agencies would definitely have to be called into this type of situation if it involved more then just the employees of the organization in question.

9. Stop the Source of the Spill

The source of the spill should be stopped if it can be done safely. This is commonly done by plugging a hole or uprighting a drum, for example. Proper PPE and the buddy system should always be used, no matter how small the source or leak.

10. Relaying Information to Employees

The supervisors should relay factual information to their employees in an attempt to relieve their anxiety. Depending upon the seriousness of the emergency, some or all of the employees should be allowed to go home for the remainder of their shift. If this is done, names and destination of the released employees must be documented by the supervisors.

11. Clean Up the Spill

Again, if it can be done safely, the spill should be cleaned up. The ERT must be familiar with the substance and have the proper training, PPE and cleanup equipment before attempting to do this. Usually a well-trained ERT can add absorbent or neutralizing materials to a spill if they know the spilled chemical. If there is any question a professional spill clean-up vendor should be called. The cleaned-up material should be treated as hazardous waste unless it can be shown otherwise, such as adequately neutralized.

12. Soil and Water

If a spill is headed for soil, a surface water body, storm drain or sewer, it should be stopped if it can be done safely. If the spill or other emergency has had an obvious or probable impact on soil or water, extra considerations must be given. An initial assessment of impact would be to determine whether more than the reportable quantity (RQ) of the chemical has entered the soil or water. An RQ is an amount specified by the regulatory authority and published in their references. If the RQ has been exceeded, the spill must be reported and it may be necessary to do a much more detailed assessment and a costly cleanup. All these activities must be reported and closely coordinated with the applicable regulatory agencies.

13. Reporting

The spill should be reported to the applicable agencies if it has exceeded the RQ or if other serious considerations have developed. This reporting must be done within prescribed time limits, depending on the agency, or fines may be assessed.

14. Re-Entry of Building by Employees

The ERT incident commander will determine (with assistance of others) and announce when the building or area is safe to enter. No one else, without exception, should allow people back into the area.

15. Closing Meeting

The ERT, management representatives, environmental/health/safety representatives and agencies involved should have a meeting after the incident is over to discuss problems, assess responsiveness to the emergency and corrective measures to minimize future occurrence. Certain results of the meeting should be relayed to the affected employee population to help relieve anxiety.

Disaster Recovery Plan

Once the immediate crisis is over, the disaster recovery plan is initiated if the business is not operating. This normally occurs after a major natural or man-made disaster. For example, if a spill, fire, tornado, earthquake or other disaster has destroyed a building, it is important that a disaster recovery plan be available to help in the recovery. Otherwise, loss of time in recovery will cost the organization production time. Financially, this can be devastating, especially if the facility or building was key to the success of the operation.

Up-Front Activities

1. Assembling the Disaster Recovery Team

In order to prepare a meaningful disaster recovery plan, a team should be assembled. After they have prepared the plan, they should meet thereafter at least once a year or in the event of a disaster. The members would include the ERT plus representatives from departments such as operations, management information systems (MIS), production, materials, facilities, environmental/health/safety, security, sales, engineering and quality.

2.   Identification of Site Resources

An inventory of critical operations and resources should be made. In case the site is destroyed in part or in total, this inventory will suggest what needs to be quickly replaced. The inventory should include a description of people, files, products and raw materials and can probably be assembled in large part by using current documents. This inventory is also important in terms of insurance coverage.

3.   Assessment of Potential Impact

An assessment of what might happen to each of the critical resources identified in step 2 must be made for possible disaster scenarios. This will help determine where backup is needed. The assessment can be in a variety of formats and Table 13–1 presents one possibility. The ranking shown is based on probability. For the site used as an example in Table 13–1, a chemical spill has the greatest potential of occurring.

4.   Potential Impact Minimization Strategies

Based on steps 2 and 3, a potential impact minimization strategy would next be prepared for the resources identified that are critical and have a high likelihood of being impacted or destroyed. For example, this may include more training, backup files and operations at other locations, secondary or tertiary containment around some chemicals and wastes, upgrading the ER plan, earthquake bracing, and additional fire suppression systems (sprinklers, hose reels, fire extinguishers).

5.   Recovery Strategy

The preceding potential impact minimization steps are to be done before the disaster. If they are done well, the business can quickly resume operation in a cost-effective manner after the disaster. It is impossible to avoid all impacts of a disaster and certain impacts cannot be minimized or avoided if a severe disaster occurs. In these situations all the organization can do is have some recovery strategies ready and then pick up the pieces the best it can, such as after a wide-area, large earthquake or war. The recovery strategy would be specific to the situation.

Table 13–1 Assessment of Potential Impact (Example)

Probability/ Likelihood	Disaster	Probable Level of Impact to		
		Environment	Employees	Operation
10	Chemical spill	10	8	4
9	Severe weather	4	6	8
8	Water damage	3	5	8
7	Fire	2	10	10
6	Explosion	5	9	9
5	Earthquake	6	8	9
4	Lightning	3	7	8
3	Terrorist activity	1	6	7
2	Airplane crash	2	7	8
1	Nuclear war	10	10	10

Note: 10 = High probability or high potential impact. 1 = Low probability or low potential impact.

6. Phone Numbers and Contacts

There should be even more emergency numbers in the disaster recovery plan than in the emergency plan. Emergency phone numbers need to be documented in the disaster recovery plan. In addition to the ones already identified in the ER plan, the following should be included:

- Property owner
- Civil defense
- Upper management

7. Routine Inspections

The organization's status, resources and supplies in terms of disaster recovery are going to vary greatly from day to day. Therefore the company resources and disaster recovery supplies need to be frequently inspected. As the resources change the disaster recovery plan should be upgraded. It is recommended that this update activity be done at least once a quarter.

8. Disaster Recovery Headquarters

If the entire operation is located in one building, an off-site disaster headquarters location should be established. This could be in another company facility as long as it is located a few miles from the site in question. It is not a good idea to put the disaster recovery headquarters in the corporate headquarters, as a significant event could wipe out both at one time.

The disaster recovery headquarters needs to be as sophisticated as the organization it is supporting. If the organization is small or not dependent upon complicated MIS-type systems, it may be necessary only to have some key backup files at the off-site disaster recovery headquarters. On the other hand, if the organization is extremely large or complicated the disaster recovery headquarters may need to be on the other end of the spectrum and look more like a command control center. In this case there should be backup files, phones, computer systems, backup power, food and water, first aid supplies, office supplies and sleeping facilities for a few key employees.

9. Preventative Maintenance

If production and environmental control equipment is maintained it will make it through a disaster with less impact to the operation and environment. Poorly maintained equipment, tanks and piping will be the first things to go when a disaster strikes. Most facility departments already have these items on preventative maintenance schedules, so that normal production can occur. Therefore it may only be necessary to verify that the schedules exist and that the maintenance frequency is adequate.

10. Backup of Computer Files and Systems

Critical data stored in computer systems should be backed up and stored off-site on a weekly basis. Also major operational software systems essential to production should be able to function at other locations, in addition to the one which may be destroyed. Critical information stored on floppies should also be transferred to the off-site location at set frequencies.

11. Backup of Paper Files

Paper files which are critical to the operation should be copied and filed at a backup location. An alternative would be to transfer the important information to hard drives, floppies or microfiche and store in a fire-resistant safe. This process can be facilitated by using a scanner to transfer the information into the computer.

12. Communications

Phone systems may go down during a disaster and severely cripple the operation's recovery. Backup capability or standby power is advised. For example, cellular phones and two-way radios are possibilities.

13. Employee Supplies

Some supplies should be purchased, before a disaster, for the health and safety of the employees who are unable to return to their homes. These might include water, blankets, flashlights, tools and food. There is usually much discussion about the proper amount of supplies to have on hand especially since some, like water and food have a limited shelf life. The author recommends having enough water and food for 75 percent of the employee population for a minimum of three days.

14. Environmental Protection Supplies

In addition to the spill cleanup supplies already mentioned, it would be a good idea to have some environmental protection supplies oriented toward minimizing environmental impacts during a disaster. This especially applies to operations that use or store large volumes of chemicals or hazardous waste. For example, depending on the operation, having backup drums and pumps in case large tanks are destroyed during a disaster is recommended.

A Risk Management and Prevention Plan (RMPP) is required by some regulatory agencies (California) for certain operations that store large amounts of toxic or extremely hazardous materials, such as chlorine and sulfuric acid. This type of assessment and plan is oriented toward protection of the off-site public and environment and may even require items like backup drums and pumps. If an organization already has a RMPP or similar plan it should be incorporated into the disaster recovery plan.

15. Facility Drawings

All important facility drawings should be assembled and stored at the off-site disaster headquarters. If this becomes too cost prohibitive, then at least the drawings showing utilities, tanks, piping, chemical and hazardous waste storage should be duplicated and stored off-site.

16. Copying and Distribution of the Plan

For obvious reasons it is important to prepare and distribute the plan *before* a disaster occurs. This entire disaster recovery section could serve as a generic outline for the plan and then site-specific information could be added. Once the plan has been completed, it should be given to the ERT, the disaster recovery team (DRT), the security command post, site environmental/health/safety representatives, security and upper management. The plan should be updated at least once a year or sooner if major changes have occurred.

During and Immediately Following a Disaster

After the ERT has the immediate crisis under control as specified in the ER plan, the following activities should occur. These would be considered the recovery actions specified in the preceding recovery strategy step.

1. Convene DRT

The ERT would already be assembled and have addressed the immediate health and safety issues and now transition into a disaster recovery team (DRT). Additional members would be added at this time and include employees from MIS, production, materials, operations and finance.

2. Inspection of the Area

The DRT would do an audit for safety hazards and, if any were found, notify employees to stay out. This would be a second safety audit since the ERT would have made an inspection earlier in the process. A third inspection of the area would also be made and a business damage assessment completed. This should include photos, documented amount of dollars required to come back into operation and recommendations.

3. Employee Needs

Even if the immediate safety needs of employees have already been addressed their longer-term needs should now be considered. This may include providing information to families or help in locating families. Employees may have other needs during and after significant disasters. For example, if a large earthquake has occurred and some of the employees can't return home, food, water, blankets and temporary shelter may also be needed.

4. Insurance Companies

The property insurance representative should be called and immediately visit the site. This individual may recommend companies that can assist in the recovery. They should be called early, before repairs start, so that the proper correction occurs and maximum coverage will be provided. Sometimes urgent repairs will have to start quickly, even before the insurance representative arrives.

5. Skills Bank

All employees who are able should report to an office that has been set up to match their skills to disaster recovery jobs they can perform safely. This not only helps the organization but also helps some of the employees deal with the disaster better since they will feel productive.

6. Reestablish Utilities

During the disaster some of the utilities may have shut down by accident or by design. The DRT should work with the gas, electric, water and sewer utilities to restore service. If chemicals are present, it is important to get the electricity on first so that ventilation systems can start clearing vapors from the area. Care must be exercised, however, since start-up of the power may result in sparks and ignition of vapors from spilled volatile chemicals and fuels.

7. Reestablish Communications

Close work with the telephone company may be needed to reestablish phone lines. In the interim it may be necessary to use cellular phones, radios or other forms of communication. Once the main telephone system is back in operation a hotline should be established to answer employee and public questions and concerns.

8. Facility Repair

The DRT should help relocate the operation, if needed, to an alternate space and/or start restoring the damaged facility. The first systems that should probably be fixed are the ventilation and fire protection systems, followed by the security command post operation. Significant damage to the facility may necessitate a temporary location. The location could be in an undamaged area or completely off-site if necessary.

If a fire occurred, the fire suppression system probably left everything wet. Moist equipment and supplies should be removed or dried promptly to prevent rust, mildew and health problems. Water-damaged documents need to be freeze-dried as soon as possible. Heat-damaged equipment and insulation should be repaired or replaced. Some smoke damage may also have occurred which could lead to product contamination or corrosion if it is not corrected.

9. Inspection and Repair of Chemical and Hazardous Waste Storage Structures

Assuming chemical and hazardous waste leaks and spills have been addressed during earlier ERT activities, it is now time to verify that the chemical and waste systems are all sound and not impacting the environment in a less obvious way. If a potential problem is noted, the system must be corrected immediately. Underground storage tanks and pipelines are high-potential problem areas and are also extremely hard to inspect. Because of this a tank tightness testing contractor may have to be called.

10. Reestablish MIS/Computer Systems

Most operations are totally dependent upon their computer systems, therefore the critical MIS functions must be repaired or relocated quickly. The backup system that supports the customers should be one of the first to be reestablished. If the backup systems were established prior to the disaster, this will be a manageable task now.

11. Replacement of Critical Files

Any essential files that may have been destroyed should be recreated by off-site files. This is especially important for certain environmental, health, safety, customer and personnel files. If floppies or microfiche were made, this task would not be as involved.

12. Reestablish Financial and Human Resource Systems

It may be necessary to shift certain financial and HR systems, such as compensation and benefit administration, to another site for a period of time. The logical location is where backup records are kept. Either hard copies, microfiche or electronic storage would be acceptable. It was suggested earlier that key financial records should be copied and stored off-site. Some organizations do this every 24 hours. If records were copied and stored, they will be available now to help the operation reestablish itself.

13. Dealing with the Media

All contacts from the media should be directed to the site's public relations manager. No other employee should make a statement. Hopefully any media releases will help gain support and assistance for those impacted by the disaster. Sometimes, however, the media coverage just adds chaos and emotional distress.

Changes/Continual Improvement

The ER and DR plans must be living documents or ever changing. This is especially important in terms of the names of the members on the teams and other resources inside and outside of the organization. The names must be kept current or the plans will be ineffective. Whenever a new person is added or a member dropped, there must be a mechanism to correct the plan in a cost-effective manner.

Chapter 14
Audits, Reviews and Verfications

Introduction

As involved as the environmental management system is, it is highly probable that certain components will not function according to plan. It is therefore necessary to audit or review the system in order to pinpoint those components that need to be corrected, adjusted or upgraded. In general, all operations having a potential for adversely impacting the environment, either physically or through chemicals and hazardous waste, should be audited. Not only should the operational activities be audited, but any paperwork that deals with the environment, such as manifests, should be audited as well. A goal should be to continuously audit systems instead of just doing a compliance snapshot.

Audits can be organized by area or by activity. For example, some organizations will do an audit once a year in each geographical area or on a random basis. Other organizations will do audits of certain operations once a year, regardless of the geographical distribution of the organization. Whichever way provides the most meaningful data is acceptable.

One purpose of the procedure presented in this chapter is to assess progress being made in terms of the policy, objectives and targets. If progress is not occurring at an acceptable pace, a correction and follow-up system should be implemented. Audits, assessments and corrections are also components of the systems presented in Chapter 15.

Frequency and Personnel Involved with Audits

Audits should be done as frequently as necessary to ensure compliance with regulations, policy, targets, certifications and business considerations. If an operation has the potential for significant environmental impact, audits should be made more frequently. Some regulations and permits require inspections or audits at set frequencies, such as the weekly hazardous waste storage area audits according to the Resource Conservation and Recovery Act. Operations that have a high environmental sensitivity or past problems should be audited more frequently.

Table 14–1   Inspections and Audits

Auditors	Subject Area	Frequency
All employees	Immediate work area	Daily
Managers and supervisors	Area of responsibility	Weekly
Safety committee	25 percent of the facility	Monthly
On-site environmental and safety representative	Entire facility	Quarterly
Corporate environmental and safety representative	Entire facility	Yearly
Outside independent auditor	Entire facility	Every two years

Table 14–1 shows a recommendation for audits ranging from daily to every two years. Every employee should audit their own work area daily. The manager and supervisor should audit their area of responsibility once a week. The safety committee should audit a portion of the facility once a month. The site environmental staff should audit the site once a quarter. The corporate environmental staff should audit all sites at least

once a year. Lastly, an outside independent auditor should audit the sites once every two years. This may sound like a lot of audits to some, however, it is impossible for any one person to see all the active or potential problems, especially in a complicated industrial setting.

The audit team should be made up of various types of individuals. The team should have at least one person knowledgeable in environmental regulations specific to the site location. One person who knows the technical process, and one who knows the facility well should also be on the team. If possible, this team should have both field and corporate representation.

Audit Procedure/Compliance Verification Procedure

 There are numerous steps in the audit process. Depending on the complexity of the site, additional or fewer audit steps might be appropriate. Regardless of the level of complexity, the following would be considered the basic steps:

1. Initial Contact

Diplomatically introduce the concept of the audit to the key site representatives and ask them to send up-front information. Some people react negatively to the word audit, so it might be best to refer to it as an assessment or something less onerous. Site employees should know about the planned visit long before the actual audit. This usually results in them correcting many items before the audit, and being more organized and helpful. As long as comprehensive correction is made, it really doesn't matter whether it occurs before or after the audit.

2. Schedule and Purpose

Schedule the audit date with the key site representatives and alert them to the purpose of the visit. Try to accommodate their production needs. Point out that one purpose of the assessment is to prevent fines and agency mandated shutdown of the operation due to regulatory noncompliance.

3. Pre-Audit Conference

Upon arriving at the site, have a pre-audit conference to go over the plans for the day. All questions that the operation personnel ask should be adequately addressed. An attempt to solicit and cover special requests made by site personnel during the normal audit should be made.

4. Review the Audit Checklist

A list of what to look for is in Appendix D. The list should be reviewed before each audit since it is impossible for the auditor to remember all issues that should be checked out. An actual audit list would be two or three times as long as the one shown in Appendix D since it should also include local regulatory issues and organization specific policies.

5. Site Inspection

The next step is to walk the site and note actions for correction. Note the items in a factual way. Also note items in a way that minimizes liabilities, if possible. For example, a note that could create liability problems might be: "A hazardous waste sign is missing at the drum storage area which is a potential fine." Instead record the same observation in a more positive way with fewer liability problems. For example, "Hazardous waste signs are required at hazardous waste storage areas. Verify that one is in place."

6. Post-audit Meeting

A post-audit meeting should be held to discuss initial findings. It is usually not a good idea for the auditors to insist on agreement concerning corrective actions on the spot. This applies whether the auditor is internal or external to the organization. To expect immediate agreement and commitment to corrective action is not reasonable and might even damage the relationship with the site operating personnel. In addition the first corrective measure to come to mind might not be the most cost-effective.

7. Analyze Findings

In order to suggest meaningful and realistic corrections the auditor should analyze the audit findings and research applicable regulations. Corrective actions that are cost-effective should be suggested. Make sure that the corrective actions are really required by regulations, company policy standards or important to employee health and safety and the environment.

8. Enter Data

Entry of the audit findings into the database should be made along with suggested corrections and if possible a reference to a specific code, standard or regulation. Give a draft to the site individuals. There are numerous database management software systems available. Some, however, are limited in capacity and sorting and analysis ability. Therefore these functions should be kept in mind when purchasing hardware and software.

9. Transmit Audit

The actual audit should be sent out to the lead site environmental, health and safety individual at the site. All other individuals who should know that an audit has occurred, such as upper management, should be copied on the cover memo. Only individuals with a need to know should receive the audit or liability problems will be created. If the actual audit goes to upper management before the site individuals are given a chance to fix the problems, resentment may be created. In addition, it is not to the best interest of upper management to concern themselves with details, such as a missing label, unless it is not corrected in a reasonable time frame.

10. Follow-up

A corrective action reminder must be transmitted if the site has not responded with a correction. It may be necessary to do this several times until all open audit items are closed. Follow-up is when most audit systems fall apart. Usually more than enough corrective actions are identified, with only a portion of them being completed.

11. Continual Improvement of Audit System

It is important to adjust the audit checklist to ensure that it is meaningful and comprehensive. The audit team composition, method of reporting and tracking corrective actions as well as other components of the audit system should be continually reviewed and upgraded.

Corrective Action Procedure for Noncompliant Items

Correcting noted problems is such an important part of environmental management that some of the suggestions presented in earlier sections will be repeated here. When action items are noted during an audit they should be corrected quickly. An organization should not create "smoking guns" which are noncorrected audit, items noted in the file. In order to make sure correction is done quickly and efficiently the following procedure is suggested.

1. Identification

The first obvious step is to identify the problems that need correction. It is better from a liability standpoint to record the issues as improvements or corrective actions in place of a list of noncompliant items.

2. Method of Correction

The suggested corrective actions may simply be one sentence or a detailed action plan. Most individuals being audited would appreciate some feasible suggestions, not just a list of problems.

3. Transmittal

Send the corrective action request to the party who is responsible for implementation of the corrections. There may be many people who actually correct a finding; therefore one person with overall responsibility for correction should receive the noncompliant items notice. There must also be someone who has the authority to insist that corrective actions are completed quickly.

4. Follow-up

Follow up to ensure that the correction has been completed. Depending on the items, this may require another site visit, a call, and/or several memo reminders. The frequency of the follow-ups is determined by the seriousness of the deficiency. An example of an audit follow-up form is shown on Figure 14–1. This form should be sent to the site at least once a quarter until all the audit items have been corrected.

ACTION ITEM TRACKING SYSTEM

LOCATION/SITE _____

YEAR OF OBSERVATION, INSPECTION OR AUDIT _____

INDIVIDUAL(S) MAKING OBSERVATION _____

Item, Issue or Problem	Specific Location, Issue or Problem	Person Responsible	Date Action Identified	Completion Date: Proposed/ Actual

Confidential

Figure 14–1

5. Continuous Improvement

It may be necessary to change an overall system or procedure that may affect several sites. The audit procedure itself may need to be changed. Continuous improvement is important here as well as in most other aspects of environmental management.

Audit Documentation

It is important to make sure that each site has the ability to document its progress and results in terms of the audits. Much data that is generated is the result of an audit recommendation. For example, monitoring and other test data are often specified in an audit. In order to complete the loop or close an open audit item, the audit documentation must be in the file. Since the audit documentation is so important, quality control is necessary. The documentation needs to follow standard protocols presented in Chapter 16.

Analysis of Audit Data

How can an organization know how well they are doing if they are not "checking their pulse" or measuring themselves against some standards or goals? Waste minimization data, injury data and emission data must be collected from audits and compared to goals and standards. To efficiently accomplish this, computer software systems should be in place for data storage and analysis. This topic is covered in greater detail in the next chapter. In general, however, additional meaning can be extracted from the data if it is entered into software systems that have the capability to perform various calculations.

Audit Reporting Procedure and Management Review

Audit findings should be presented to senior management and to employees that can help in the corrections. As with the other presentations to senior management, audit findings should be illustrated in a clear and descriptive way. One possible way of showing audit findings is illustrated in Figure 14–2. There are an infinite number of ways of illustrating audit findings. As mentioned earlier, senior management should be shown the big picture and problem areas that are not being fixed.

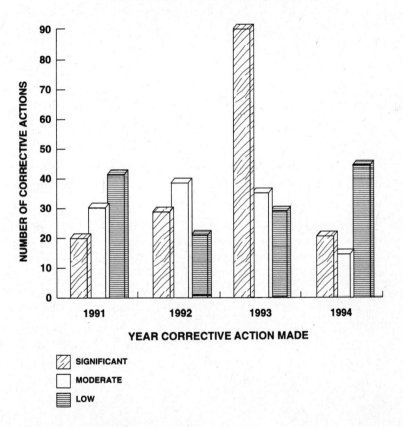

NUMBER OF CORRECTIVE ACTIONS MADE EACH YEAR AND LEVEL OF SIGNIFICANCE (EXAMPLE)

Figure 14–2

The audit results must be relayed to relevant management individuals in a meaningful way. One way of making the results useful to management is to relate them to the objectives and targets of which they should already be aware. Based upon this type of presentation management will be able to implement positive change.

The findings and recommended corrective actions from audits should be relayed in writing to the site environmental manager, and all others that should be aware that an audit took place would only be copied on the cover memo. Usually it is politically advantageous to give the site environmental manager an opportunity to correct the items before senior management becomes involved. If the site environmental manager facilitates correction of the significant findings promptly and completely, it would not be necessary to give senior management the complete audit. If at least the cover memo is sent to senior management, it will let them know that an audit has been done and a phone number to call if they want a copy. The memo should also state that all significant discrepancies uncovered by the audit have been corrected by on-site personnel, if this is the case. Senior management should be notified when certain audit items are not addressed.

Site management should use the audit findings to their full extent. This is valuable information that can help the organization improve. It is also information that can lead to liability problems if it is not acted upon. It may be advisable to modify targets, objectives and policies based on the audits. When and if this is done, the site management must be involved.

Continual Improvement

After the programs, process and plans discussed in this chapter are implemented they must be continually improved. The improvement would be based on data obtained, audit results and other types of feedback available. Reengineering and benchmarking in reference to audits are aspects of continual improvement.

Chapter 15
Data Collection and Handling

Introduction

It is important to be able to track key variables that show impact to the environment, compliance with regulations and permits, success of the procedures and programs and whether the policy, targets and objectives are being met. To properly manage the operation and the environment, certain components need to be monitored or tested. This allows the environmental manager to make meaningful changes, corrections or adjustments. If this is not done, a considerable waste of time and money is possible with little benefit to the environment. The audits just discussed generate data. Remediation studies, air and water monitoring, equipment measurements and waste testing also result in the generation of data. How the data is collected and analyzed will determine its usefulness.

Determination of Data Needs

Most processes that use hazardous materials or generate discharges, emissions or wastes should be monitored. The variables measured and frequency vary significantly for different processes. Permits and regulatory requirements also dictate the monitoring that must occur.

Data or monitoring requirements are related to environmental and safety needs. Environmental monitoring should occur where there is possible impact to air, surface water, wildlife, soil, land or ground water. This monitoring would be of either the discharge itself and/or the receiving area of the ecosystem. Safety-oriented monitoring should be done where there is possible impact on employees. As with environmental monitoring the samples would be of the emission and/or the transport medium to the employee, such as air.

Data Collection Procedure

The EPA and other agencies have published procedures for sample and data collection. These procedures specify accepted methods. Due to the wide variety of types of data it is impossible to summarize them all here. The following generic principles, however, could help an organization start to develop their own specific data collection procedure to comply with ISO 14000.

1. Sample Collection

The most cost-effective, safe and accurate method to collect the samples should be specified. Things to consider include whether it is a grab or composite sample, appropriate location of the sample, volume, sample equipment needed and numerous other factors.

Most samples will be either composite or grab. Composite samples are taken over time and are usually for lesser toxic substances. Grab samples, such as for hazardous substances, are collected at one moment in time. A grab sample may not be as accurate; however, it minimizes possible exposure time for the sampler.

2. Sample Containers

The samples should be placed in the proper sample containers. Care is needed since certain substances can dissolve glass or plastic containers. Breakage during transportation is also a consideration. Therefore, if plastic is not dissolved by the sample, it is the preferable material to use.

3. Transport of the Samples

Usually the samples should be transported quickly and in containers that will not break. In some cases this must also be done with preservatives or on ice. For example, volatile organic compounds may off-gas during transportation resulting in inaccurate results if proper procedures are not followed.

4. Field Analysis

A quick and usually inaccurate result can often be obtained by using field analysis methods. These may include test strips such as pH paper, field meters such as those used to measure conductivity and test kits such as the HAZCAT. One HAZCAT kit on the market will allow a field determination of the major hazard classes such as flammability. Another more sophisticated HAZCAT will allow a rough determination of approximately 20 or 30 different substances such as phenols. Even if field analysis is not too accurate, it still gives the environmental manager quick information that is useful in several situations such as in emergency cleanup and further sampling locations.

5. Laboratory Analysis

The samples must be analyzed in a certified laboratory. Even if some analysis can be done using field instrumentation, a laboratory is usually used for greater accuracy. Certified laboratories must have detailed quality control (QC) systems. Many different types of laboratory equipment, such as atomic absorption, inductively coupled plasma emission spectrometry, gas chromatography and high-performance liquid chromatography are available to analyze environmental samples.

The above is a gross oversimplification of a complex science. Whole books have been written on data collection and analysis with entire chapters dealing entirely with quality control. Because of this complexity, most organizations use consultants and laboratories to perform the data collection and analysis for them. Data collected and analyzed in this way will also stand up better in court.

Data Interpretation Procedure

The interpretative processes basically take data and extract new meaning from the numbers. It is not enough to say there are 5 ppm of lead. Does it comply with the regulations? Does it act synergistically with some other chemical? How significant is its impact on employees and the environment? Data interpretation helps the environmental manager answer these and other important questions.

Probably the most widely used interpretative procedure is called the statistical survey research method. In this method the data is analyzed using averages, means, standard deviations and other statistics. This would probably be a good first step in extracting new meaning from the data.

The descriptive survey research method is also widely used to analyze scientific data. In this method the data is compared to all other forms of related information that can be obtained by surveying relevant documents such as: environmental impact statements, regulations, permit applications, agency files and a wide range of technical publications.

Monitoring Equipment Maintenance

All monitoring equipment needs to be maintained. This applies whether the organization or a consultant performs the monitoring. Field sample equipment and field and lab analysis equipment must all be properly maintained so that the data is accurate. Maintenance activities such as cleaning, calibration, repair, replacement and upgrade should be recorded.

Data Management Tools

There are numerous data management tools and systems available on the market today. Many of these involve high-powered software programs. For example, MSDS information can be loaded into a chemical database and combined with chemical inventory information. This allows the environmental manager to track chemical usage and general information more accurately. Recycle data, injury data, discharge data and most other environmental information take on new meaning if they are managed in this way.

Use of the Data

In general, data should provide new insight and be the basis for making meaningful and positive change. For example, the data collected can be used to adjust objectives and targets and to improve processes and systems. Identification of corrective actions, implementation of controls and preventative actions would result. This requires adjustment based upon the data. The data should also be used to determine areas of non-compliance or recorded in order to show compliance. This will help if fines or litigation develop. The data, or a summary of it, should be reported to senior management.

Reporting of the Data to Senior Management

The reporting of data to senior management is of special importance. Most organizations do this in a variety of ways and without any set procedure. Problematic data is commonly relayed to senior management quickly and verbally. The rest of the data is usually not relayed at all or is inconsistently reported. Senior management should get good and bad data consistently and in an understandable way so that they can run the organization in the appropriate manner. The following procedure is recommended to accomplish this goal.

1. Needs Determination

Determine the types of data senior management needs. For example, this may include changes in discharge quality, potential litigation or fines, significant new expenses and public and customer opinions.

2. Presentation Package

The data should be packaged in an accurate and quickly understandable format. It would also be helpful to senior management to include the targets/objectives and regulatory limits so that they can put the data into proper perspective.

3. Transmittal of Data

Send the data at the appropriate frequency that will make an impact. For example, problematic data should be sent when it is known and other important data at least once a quarter. A face-to-face presentation of the data is much better than mailing it so that senior management can ask questions and you can provide interpretations.

4. Recommendations

In order to come to a closure with senior management concerning the presentation of the data, it is important to conclude by making recommendation(s) for improvement and asking for management approval or suggestions.

5. Follow-up

Sometimes senior management cannot or will not give an immediate approval of your suggested course of action. If this is the case, it is important to remind them that you are waiting for their approval to proceed with your course of action. They are busy running the business and may need several reminders. They should also get updates on progress after giving approval.

Chapter 16
Records and Document Control

Introduction

Many regulations require that environmental documents be retained for specified periods of time. This, plus the fact that numerous lawsuits occur in the environmental field, makes record and document control very important.

All the environmental management procedures, plans, targets and almost everything discussed so far would be considered documents requiring control. All these documents, elements or procedures should be assembled in one place so that their interaction or relationship will be obvious. An organization has the option of assembling all of these documents into one environmental management manual or leaving them in their present form.

In general, the types of documents that should be collected and retained are those that are meaningful in terms of environmental management. They should include a description of the core elements of the environmental management system, such as targets and objectives. Second, the processes, procedures and other systems should be collected and retained. Regulatory and any other useful sources of information, such as manifests, must be retained.

Handling of Records

1. Identification of Appropriate Documents for Retention

A first step in document control is a thorough identification of what needs to be retained for regulatory and business reasons. Table 16–1 lists many of the environmental documents that should be retained and an estimate of the amount of time to keep each document. For example, this list shows that the manifest is to be kept for three years. RCRA requires this, and an organization can be fined up to $25,000 per day if it is not done. Many of the other dates shown in Table 16–1 are not required by regulation but make good business sense. For example, if a fine is assessed or litigation develops it may be important to have copies of these documents.

2. Collection of Documents

One obvious way to collect documents is to instruct individuals who may be involved with the document to submit a copy to you. If the person is busy, this may not be too reliable, and it is, therefore, better to either pick one up yourself or be involved with the original document and make your own copy. Close supervision to assure compliance is needed.

3. Indexing of Documents

Table 16–1 is arranged to illustrate one way that environmental documents can be indexed. There are, however, an infinite number of other ways to organize the documents. Whichever way is selected, it should be computer-based to permit easier searching by subject.

4. Filing and Storing of Documents

Environmental files need to be well organized, kept up to date and locked. Part of the reason for this is the high number of environmental lawsuits. It is also a good idea for the documents to be retained in either a fire-rated room or file cabinets or a backup copy kept at another location. If the volume of files exceeds the space available, it may be necessary to have a set of active file cabinets on-site and inactive files off-site.

5. Removal of Obsolete Files

Once the time listed on Table 16-1 is reached, it is important to promptly destroy the outdated files for two reasons. First, old paperwork can become a nightmare during litigation, since it would probably have to be produced (delivered to the court) if it still exists. This means every draft and old document, even if they are no longer relevant. Also, most

Table 16–1   Records to Maintain and Duration

Corporate Records/Files

- Waste minimization—10 years

- Environmental policy—while active, +5 years

- Litigation, fines, orders—6 years after resolution

- Internal audit files—while items are uncorrected, +1 year

- Contractor information and audits—2 years

- SARA records—5 years

- Biennial reports—3 years

- Manifests—3 years

- Details of nonconformance and corrective actions—10 years

On-site Records/files

- Training records—10 years

- Monitoring data—30 years

- Internal audits—while items are uncorrected, +1 year

- Failures, incidents, complaints and follow-up actions—30 years

- Maintenance records—10 years

- Chemical inventory reports—30 years

- Hazard communication documents—30 years

- MSDS—10 years

- Permits—while active

- Underground storage tank records—10 years after closure

organizations have limited space and purging old files will free-up room for more important documents. For the removal of obsolete files to be accomplished promptly and efficiently, the computer-based indexing system mentioned above will be of value.

Document Control Specifications

The same document control specifications required for ISO 9000 certification apply here as well (refer to Table 16–2) to both paper and electronic forms. For example, each page must be numbered (such as page 1 of 3). All documents must have issue date, retention time and revision numbers. If the document is related to another it must be cross-referenced. Since the specifications are the same, they will not be repeated here.

Table 16–2   Minimum Document Control Standards

- Effective date
- Responsibilities
- Approval signature line
- Title of document
- Procedure, product or part number
- Effective date
- Revision number
- Purpose of the document (work instruction, procedure, etc.)
- Scope/applicability of the document
- References
- Definitions
- Page numbering (page 1 of 4, 2 of 4, etc.)
- Procedural steps

Preparation and Distribution of Documents

A procedure for preparing and distributing documents should be followed. As an illustration, the following general steps apply to a standard operating procedure binder for environmental, health and safety. The most important goal of this entire process is controlled preparation and distribution of the environmental procedures to the proper individuals in a cost-effective manner.

Entire Binder Controlled

1. Preparation

The entire binder will be prepared by the corporate environmental, health and safety (EHS) department with input from others as appropriate.

2. Distribution

One numbered and recorded copy will be given to the lead site individual, the senior environmental, health and safety individual, the safety committee chair and the training coordinator.

3. Copy Requests

All requests for a copy should be directed to the corporate environmental, health and safety department.

4. Verification of Receipt

After the individual receives a copy of the binder they must send back the receipt sheet verifying that they received the binder, have completed a review, understand its contents and will comply to the best of their ability.

Single Procedures—Controlled

1. Revision or Preparation

After a single procedure has been revised or prepared for the first time it must be forwarded to the corporate environmental, health and safety department for checking and inclusion into the controlled binder.

2. Distribution

The corporate environmental, health and safety department will forward approved individual procedures to each binder holder, along with an updated index.

Single Procedures—Uncontrolled

It may not be cost-effective or necessary for all employees to have an entire controlled binder of procedures. For example, certain administrative employees may need to understand only a few procedures. In these case(s) the following procedures should be followed:

1. Copies

The site environmental, health and safety individual(s), supervisor(s) or any other site individual(s) who has a controlled binder may make uncontrolled copies of single procedures, as needed.

2. Stamping

Before the copy is given out, it must be stamped "Uncontrolled." The recipient should be told that they are not on distribution for updates, since only controlled binder holders get revisions and updates automatically.

General Distribution Guidelines

Before documents are distributed, a determination of whom should receive them must be made. Figure 16–1 shows a distribution matrix for one type of document, environmental, health and safety standard operating procedures (SOPs). Once the matrix is completed, it will show that certain populations of employees should get only certain SOPs and some employees should get all SOPs.

As with document control, the distribution specifications are anticipated to be the same as they were for ISO 9000. For example, controlled copies sent out are to be numbered, whereas uncontrolled copies are not numbered. Only the controlled copy holders receive updates. When certain important documents are received, a return receipt should be sent back to the distributor. Figure 16–2 shows one possible form for ensuring that documents have been received.

**MATRIX FOR ILLUSTRATING ENVIRONMENTAL,
HEALTH AND SAFETY PROCEDURE DISTRIBUTION**

Job Description	Job Code	Applicable Procedure
Environmental, health and safety		
Hazardous waste handlers		
Managers and supervisors		
ERT members		
Maintenance personnel		
Engineers		
Technicians		
Chemical handlers		
All other employees		

Figure 16–1

**ENVIRONMENTAL, HEALTH AND SAFETY
DOCUMENT RECEIPT FORM**

DATE:

TO:

FROM:

SUBJECT: Receipt of Environmental, Health and Safety Procedures

I have received the following procedure(s) and have read and understand them. I will make every effort to comply with these procedures.

Procedure(s) Name and Number_____

_____

_____

_____

_____

_____

_____

These procedures have been filed in my controlled binder number_____ , which can be found in my office (or specify other location_____).

_____

Signature

_____

Typed or Printed Name

_____

Date

Figure 16–2

Attorney-Client Privilege

Certain documents and verbal communications need special protection, not because they deal with illegal issues but because they contain inappropriate material for the courtroom. For example, some of the first paperwork or statements made concerning an environmental problem will probably be full of guesses, erroneous theories, errors, assumptions, value statements and other incorrect and misleading information. If this material is not protected under Attorney-Client Privilege, it can legally be "discovered" or brought to court and result in days of wasted time going down an incorrect path. This is not in the best interest of anyone involved.

The best way to implement Attorney-Client Privilege is by getting the attorney involved early. If the attorney requests files, samples, and other information, the material will be prepared at the request of counsel and protected under Attorney-Client Privilege. All documents generated in this way should be clearly marked Attorney-Client Privilege. The following list presents suggestions of when to consider implementing Attorney-Client Privilege protection.

1. Public Involvement in a Sensitive Issue

The public becomes involved in a sensitive internal issue which has the potential for litigation and negative publicity. This could include individual citizens, environmental and other public-interest groups.

2. Problem Employee

Negligence or criminal conduct on the part of an employee is suspected. This should be handled properly by the attorney from the start.

3. Violation

A serious violation has probably occurred and should also be handled under the direction of the attorney as soon as it is known.

4. Significant Environmental Damage

If it is probable that significant environmental damage has occurred, the attorney should be called immediately. Involving the attorney should not, however, delay action to help contain the situation or minimize the impact. For example, if a spill has occurred that involves public property, sewers, soil, storm drains or air quality, it has to be reported. Once the immediate crisis is under control, then call the attorney.

5. Significant Health and Safety Impact

When a significant health or safety impact has occurred, fast corrective action must occur, and then the lawyer must get involved. For example, when an employee ends up in intensive care because of a work related accident, the attorney should be called.

6. Law Enforcement Becomes Involved

When the police, FBI or district attorney become involved, it would probably be best to get the organization's attorney up-to-speed immediately.

7. Lawsuit

When a lawsuit has been threatened Attorney-Client Privilege should be implemented.

8. Contamination Is Found

Site investigations where air, water or soil is sampled and significant contamination is found above regulatory limits are red flags indicating legal protection may be warranted.

9. Notice of Violation

Implement Attorney-Client Privilege when a notice-of-violation or other notice of noncompliance is received

10. Contact by an Outside Attorney

If an attorney contacts the organization on behalf of someone else, it should be assumed that legal action is imminent and the organization's attorney should initiate further actions.

Chapter 17
External
Communication

Introduction

Since no organization operates in a closed system, it must communicate with the outside world. Some components of that world are customers, media and the general public. Special communication procedures are warranted for sensitive groups. Environmental information can be effectively relayed to these groups by agency reports, annual reports, industry group documents, marketing fliers, product markings and numerous other ways.

In general there are three main steps that should be taken when handling external communications. The communication coming in from an external source should be accurately recorded. After the documentation of the request has been completed, some research should be done to determine the correct answer or response. Lastly, a response should be made or an entry placed in the file why a response was not made.

Customer Communication

Some organizations receive numerous customer requests for environmental information. In the future when an organization with ISO 14001 certification receives a request, in many cases they may just have to show the certification.

Presently, this is not the case. Now numerous requests for different types of environmental information require research and unique answers. This is an extremely time-consuming process if the organization is fortunate enough to have many customers who are environmentally conscious. In this case a tracking system is needed so that all customer requests and responses are indexed by name, date and subject. It is important that one group in the organization respond to all customer requests for environmental information. This will ensure consistency and accuracy.

The organization has an obligation to notify their customers of any chemical hazards present in the product. An MSDS is the common avenue for doing this. The customer should be urged to reuse, recycle or, if all else fails, to dispose of the product in an environmentally conscious way.

Customer requests can be extremely varied. In the author's experience, however, most requests concern the organization's use of ozone depleting compounds, PBBs, lead, cadmium and recyclable packaging materials. The following list is a sample procedure for handling customer requests for environmental information.

1. Initial Receipt of Request

Whoever receives the customer inquiry should attach as much information as is readily available to the request and forward it to a central group that issues all responses. The central group is usually the corporate environmental, health and safety department. They should routinely communicate to the organization that they are to receive all customer requests for environmental information.

2. Determination of Whether a Response Is Required

The central group will determine if there is a regulatory or legal requirement to provide the information. Certain regulations, such as California's Proposition 65 (Safe Drinking Water and Toxic Enforcement Act), require notification of certain environmental, health and safety information concerning products. Legally binding contracts between the organization and the customer will sometimes require release of environmental information about the product.

3. Determination of Whether There Is a Business Reason to Respond

If there is not a legal or contractual requirement then the client coordinator will determine whether there is a business reason to respond.

Keeping a good customer happy and being responsive is one of the best business reasons for providing environmental information.

4. Preparation of Response

If the response is made, it must be accurate and consistent with responses made to other customers. The initial response should be brief in order to minimize liabilities and if this does not satisfy the customer, then more detail would be provided as a second step.

5. Research Required for Certain Responses

In some cases a response may require laboratory testing or extensive research. If this determination is made, the customer should be notified that there will be a time delay in responding to the request and a possible increase in cost of the product or service.

Media Communications

It is a good idea to have a well-established procedure for handling contacts by the media since these may be sensitive exchanges. It is in everyone's best interest that the information is reported in a factual and nonemotional way. A few suggestions follow.

1. Organization Initiated Contacts with the Media

The organization should establish contacts with the media prior to the development of a sensitive issue. This will help establish a good working relationship and should start with positive media coverage.

2. Receipt of Media Call

Whoever receives the initial call from the media should get the caller's name, number, and questions. This information would then be immediately relayed to the public relations department within the organization. Under no condition should the person answering the initial call provide answers. Even if this is the technical person, they need time to think about the answers.

3. Return of the Call

Both the public relations individual and the technical individual should return the call together. Some time should be spent before calling back, however, to determine the proper answer. The delay in returning the call should not be too great or the reporter will just print their version of the story.

4. Content of the Returned Call

All answers should be brief, factual and as positive as possible. All questions and answers should be recorded. If an answer is unknown, it is better to say so than to guess.

General Public Communications

When someone from the general public has contacted the organization, the comment should be documented. An analysis should next occur to determine whether the comment has merit and will help improve the environment. The determination of action or no action should also be recorded and relayed to the individual. Good ideas from the public should be implemented.

Chapter 18
Design for Environmental and Life-Cycle Analysis

Introduction

The term "design for environment" captures the essence of the new movement in environmental protection. It takes the old reliable, well-proven environmental control methods and concepts and combines them with new ideas that concentrate on the design phase of a product. Closely related to design for environment are the terms life-cycle analysis, sustainable development and green design. These all have slightly different meanings but also have the common element of environmental awareness during the design phase.

It may not be possible to evaluate every product, part and raw material impact. To do so might be cost-prohibitive. If this is the case, then one could lump products, parts, or raw materials into classes or groups. The class or group with the most significant predicted impact would be analyzed first followed by groups with lesser impacts, until money or time runs out.

Key Components

An inventory of impacts is the first key component of design for environment. When this is done, it is important to identify impacts from all phases of the product life cycle. Air and water impacts may occur when raw materials are produced, fabricated, manufactured and assembled into the final product. The post-production phase will also generate impacts such as when the product is used (energy consumption, etc.) and during disposal. Some feel that the greatest impacts occur during the use and disposal of most products.

171

There are many environmental issues that should be considered during the design phase of a product. For example, the ability to easily disassemble and then reuse or recycle are important environmental considerations. Incorporation of materials with no or low toxicity is also at the top of the list in terms of design for environment (NCMS, 1994).

Also important is the use of chemicals with the lowest toxicity possible in the manufacturing process. If this is not possible, then a chemical that can easily be recycled and reused should be used. If it is not possible to do this on-site, then an off-site recycler who will recycle the used chemical and return it for reuse should be employed.

Customer's Part in Design for Environment

Design for environment takes creativity, hard work and a different mind-set by not only the manufacturer but the customer as well. The customer has got to be willing to accept a product that may not be quite as high in quality. In place of perfect quality the customer would get a product with reused parts or environmentally friendly components. This may mean that a product may not be able to perform without error in exchange for one which contains recycled parts or can be easily disassembled and recycled. Everyone has got to give up a little if design for environment is going to become a way of life.

Benefits of Design for Environment

Obviously the greatest benefit of design for environment is the reduced impact to the environment. This benefits all components of the ecosystem including people. Consumer benefits are seen in cleaner air, soil and water upon which we all depend. Fewer dollars will be spent on remediation that is not always effective. Overall the quality of life will be improved.

Design for environment can also lead to long-term cost savings for a manufacturer. If some chemicals are not used during the process, there will be less hazardous waste to handle which is a cost savings. Fewer legal expenses, fines and penalties are also probable if this approach is followed.

Standardization

Several organizations are attempting to standardize design for environment or green design. For example, the National Center for Manufacturing Sciences has come up with a Green Design Advisor project that is oriented toward helping manufacturers minimize effects on the environment at the design phase. They are developing software tools and the information network necessary for this standardization to occur (NCMS, 1994). The ISO subcommittee led by the Germans and French is also involved in life-cycle analysis and design for environment standardization.

Environmental Design Database

Combined knowledge from various sectors of society will be necessary to make design for environment work. An environmental design database which compares various processes, chemicals and other materials in terms of environmental impact, cost and quality will need to be developed. This tool would be used by the new product designers to help them find ways to minimize environmental impacts. To develop this database, it will be necessary for communication networks to become established between competitors, suppliers, vendors, recyclers, universities and the government. This will allow information sharing and entry into a system that everyone can use.

Summary

Life-cycle assessment and design for environment are powerful and important concepts. The environment will definitely benefit if these procedures are followed. Care must be exercised, however, since application of this concept can become expensive, hard to do and somewhat academic without proper planning.

Chapter 19
Minimizing Discharges to Air, Water Bodies and Sewers

There are literally hundreds of specific procedures for minimizing discharges to the environment and achieving the policies, objectives, targets and regulations. Due to the variety of types of industries and locations, these procedures can only be discussed generally in this book. Numerous examples have already been presented in earlier chapters. Specific examples of procedures for minimizing discharges can be found in the 1995 book by Kuhre entitled *Practical Management of Chemicals and Hazardous Wastes*. Two chapters from this book are presented in Appendix D, "Assessment Procedures," and Appendix E, "Waste Minimization Procedures." These two chapters are so basic for minimizing discharges and obtaining certification that they are reproduced in total. A few generic procedures are summarized below:.

Procedures for Minimizing Discharges to Air, Water Bodies and Sewers

In a gross oversimplification, the operator should measure the quantity and quality of pollutants coming from each specific process, design and implement emission control that will meet regulator compliance. This would apply to point and nonpoint sources of pollutants. General procedural steps include:

1. Base Line

Base-line air or water quality measurements should be taken. This will provide valuable information for future reference. It will show progress made by the organization and might also differentiate between present pollution and past pollution.

2. Emission Estimates

Before detailed design occurs, it is important to make estimates of the probable emissions. Even if the estimates are rough they will allow discussions concerning possible control equipment and systems to begin.

3. Regulatory Requirements

A check of the applicable regulations and contacts with the regulatory authority will be necessary in order to obtain a permit to construct. This should be done very early in the process since expensive controls may be required.

4. Impact Minimization

It is important to reduce as many potential emissions during the design phase as possible through controls and waste minimization. This may cost some up-front capital; however, there should be a cost savings over the life of the operation.

5. Installation

During the installation of the process and control equipment, attention should be directed toward regulatory requirements. Permit stipulations must be followed very closely. If modifications are needed, and most complex installations need some modification, the changes must be run by the regulatory authority unless they are very minor in nature and do not change discharge points and amounts.

6. Training

Operators should be trained in process and environmental considerations. This training is necessary for employee safety as well as environmental protection. Training will be both initial and annual refreshers for most classes.

7. Permit to Operate

Once the training has been completed, the company should obtain a permit to operate and start operations. In some locations a separate type of permit is required. The regulatory agency will probably need to come out for inspections, usually after the process has operated for a period of time.

8. Sampling/Monitoring

Many regulations require the sampling of actual discharge or emissions from the process to verify regulatory compliance. At a minimum the variables specified in the permit must be assessed.

9. Upgrade Controls

Depending on the actual data obtained, it may be necessary to improve controls. This may also be just a matter of adjusting the existing control equipment. Sometimes, however, major modifications will be needed to allow the controls to remove pollutants to regulated levels.

10. Maintenance

Both the process and control equipment should be carefully maintained. This is very important since many problems are related to lack of proper maintenance. Some health, safety and environmental disasters in the past are believed to have been caused by lack of maintenance, such as in Bhopal, India.

11. Logs

It is good management to keep logs of fuel and chemical usage. Some permits may require logs, and, even if they do not, this type of information is needed for mass balance tracking.

12. Ongoing Monitoring

Ongoing monitoring must occur to ensure compliance with permits, regulations and company policies. Monitoring will allow continuous improvement to be possible. There is really no other way to accurately know that the system is operating properly.

13. Audits

Routine audits of the operation should be made and adjustment carried out to continually reduce emissions, chemicals and wastes. The audits would be done in response to special problems in addition to scheduled or routine audits for regulatory compliance.

14. Special Problems

Unauthorized discharges, spills and other problems must be reported immediately to company management and to the regulatory authority. In certain cases forms will have to be completed.

Chapter 20
Management of Nonhazardous Materials and Wastes

Since there is a shortage of landfill space, it is important to have a comprehensive procedure for minimizing waste. This applies as much to households as it does to industry. In the following procedure, paper, aluminum, glass, cardboard and plastic will be addressed. For all these materials the four Rs apply, that is, reduce, reuse, recycle and repurchase.

Reduce

If everyone tries to reduce the amount of materials they use, there will be less volume to handle. For example, if double-sided photocopying is done, less paper will have to be purchased. Use of E-mail or other electronic systems, computer files and restricted distributions will also reduce the amount of paper used.

Reuse

By reusing paper, cartons, packing and other materials, the volume of waste to be handled later will be reduced. If the material can't be reused for the original purpose because of quality reasons, perhaps it can be used for some other purpose. For example, it may not be possible to reuse a carton to ship new product to a customer; however, the carton could be reused for shipping defective parts back to a supplier. By reusing materials to the maximum extent possible, the organization is saving resources, money and valuable landfill space.

Recycle

1. Materials Premarked for Recycling

Some materials, such as plastic, have recycle symbols already printed on the new product. It may even be possible for the user organization to mark some recycle symbols on supplies and other products, in some cases, if they do not already contain a designation. These symbols help the user know what is recyclable. Care must be exercised, however, that use of the symbols is backed up with data to support the claim. There are also specific definitions and criteria for use of recycle symbols.

2. Inventory

An accounting of types and volumes of materials to be recycled should be made. This may involve collecting figures from all of the organization's sites in order to achieve economical volumes. The figures can be estimates but should be conservative. In some parts of the world, such as Germany, law requires that the manufacturer inventory and take back packing materials for recycling.

3. Contract with Recycler

An agreement should next be set up with one or more recyclers. Before signing the agreement it would be a good idea to research several vendors to make sure they are really going to recycle the waste and so that the best deal can be obtained. Some governments, such as Germany, may require the use of a specified recycler (Green Dot Program). Details such as who provides the collection containers, dates of collection and charges or credits need to be worked out and reflected in the contract. Another detail that is important to work out is whether all materials will be put into one container and sorted at the recycler facility. If space permits, it is usually better to have separate containers at the generator's location. This would mean separate containers for paper, cardboard, plastic, glass and aluminum.

4. Placement of Recycling Containers

Unfortunately, placement of the recycle container in relation to the point of generation of the waste materials plays a big part. If it is too far away or confusing to use, many people will not take the time to recycle.

The containers should be placed as close to the generators or source of the waste as possible. They should be clearly marked, especially if the materials to be recycled are to be placed into different containers.

If there is a "Take Back" type law in place, such as in Germany, a separate container may have to be set up for packing material and used product which have been returned by the customer. It is especially important that this material be accurately accounted for and proof of recycling received.

5. Pickup and Recycling by the Recycler

When the materials are picked up, it is important to obtain certain documentation. In addition to a weight receipt, some sort of certificate or notice that the material is to be recycled is important to have.

Repurchase

To complete the cycle or loop, it is important that the organization purchase recycled products, even if they cost a little more. The purchasing department may need to set up volume discounts to help reduce this cost. If more organizations would purchase postconsumer products or recycled products, it would improve their marketability and price. This is what is really needed to make recycling work.

General Procedures to Ensure that All the Programs Work

There are some common elements that will help all the reduction, reuse, recycle or repurchase programs be successful. Actually these suggestions would maximize the success of any environmental program. They include the following.

1. Employee Awareness Program and Reminders

Most people need to be "energized" about reuse and recycling and continually reminded. It is just not at the top of the priority list for many busy individuals. Reminders can be in the form of fliers, posters, verbal announcements and other means. Waste minimization contests can be valuable as a reminder to recycle.

2. Tracking and Auditing

As with most environmental programs, it is necessary to track and audit for compliance or progress. In this case the auditing could be done by the site's waste minimization committee. If the site is very small and does not have such a committee, then the site's environmental representative or someone from corporate should do the audit.

3. Targets

Each site should set up some targets along with some corporate-wide goals. As discussed previously, the targets would be actual numbers, such as pounds reduced.

Chapter 21
Site Closure Procedures

An organization must properly close down their operation in an environmentally conscious manner. This is required by law and a full disclosure must be made to the buyer of the property. Some of the procedural steps should include:

1. Inventory

An inventory should be made to assess areas requiring attention. For example, on-site raw materials, wastes, chemicals, fuels, production equipment, waste treatment equipment and electrical transformers should be itemized.

2. Phase I Environmental Assessment

A Phase I environmental assessment may be necessary and would be prepared by a consultant. This would especially apply to all manufacturing, research and distribution facilities with a history of environmental problems or where hazardous materials or hazardous waste have been stored.

3. Regulatory Requirements

An identification of local, state and federal requirements applicable to site closure should be made. For example, an EPA ID number may have to be closed out.

4. Closure Plan

Depending on the site, the Phase I environmental assessment and the regulatory requirements identified above, a written closure plan may be necessary. This plan would provide details concerning all of the actions in steps 5–13.

5. Disposition of Remaining Materials

All remaining raw materials, chemicals, fuels, waste and debris should be removed. The inventory prepared earlier should be checked off as these materials are being removed. If any of these can be sold, it would be better for the environment and the organization's liabilities. Transportation of chemicals requires a bill of lading and hazardous waste transportation requires a manifest. Both must be done by a licensed handler.

6. Cleaning of Equipment

Production equipment, storage facilities, waste treatment and all other equipment that encountered chemicals or hazardous wastes must be thoroughly cleaned. This would include piping and ducts. Status of the cleaning wastes must be determined according to the regulations and will probably be considered hazardous waste. Tests should be done on the cleaned equipment to make sure that chemicals are at an acceptable level. This level would depend on the landfill or buyer's requirements and on regulations.

7. Special Hazards

Special procedures and handling would be necessary if asbestos, PCBs, on-site disposal or other special hazards are present. Numerous regulations specify how these substances are to be handled.

8. Notification

The applicable government authorities must be notified of site closure. This may also include submittal of a closure plan and ending an EPA ID number. The notification should be done in writing.

9. Permit Review

All environmental permits should be reviewed in terms of closure and appropriate action taken. A permit might require that an area or pond be reclaimed, for example. Removal and decontamination of equipment are also required in most cases.

10. Photos

Photographs should be taken to document the condition of the interior and exterior of the site after completion of the closure activities. This is especially important if the next owner alleges that certain structures were left behind.

11. Final Site Inspection

A thorough inspection of the facility after all closure activities are completed is necessary. This will help ensure that closure has taken place in accordance with the procedures outlined above.

12. Closure Report

A report documenting the findings of the final site inspection should be made and filed as specified in step 13. All observations should be recorded in case disputes arise in the future.

13. Records

Environmental records should be collected and transferred to a secure location. Many of the records, such as manifests, must be retained for a minimum of three years. Other records which should be saved include the closure inventory, closure plan, closure report, permits, Phase I environmental assessment, purchase agreements, cleanup testing, photographs, hazardous waste training records, bills of lading, shipping papers, consignment notes, chemical purchase records, waste testing results, reports required by permits and any other environmental record that seems appropriate at the time of closure.

Chapter 22
Conclusion

There is a monumental change about to occur that is going to significantly improve the environment and affect the way that organizations do business. For the last 10 or 20 years most organizations around the world have implemented environmental controls largely because of potential fines, penalties, jail sentences, litigation and other regulatory reasons. Many of these driving forces will always be around, but now a much more important and positive reason is about to occur. This is going to result in organizations going much further to protect the environment. The reason for all this is the customer is now requiring the environment be protected to a much greater degree. This requirement is being expressed in various ways; however, very soon the most common form will be an ISO 14001 certification. When the customer wants something, it usually happens or the supplier eventually ceases to exist.

Industry may feel that an ISO 14001 certification and the new environmental management systems are going to add more cost and they are right, at least initially. Some upfront cost will undoubtedly be encountered; however, over the long run there should be cost savings due to avoided cleanups, fines and other costs related to environmental problems. These problem-related costs are the ones that can bankrupt an organization, not proactive ISO 14000 type expenses.

One could write about the environmental management systems almost indefinitely due to the breadth of the field. When should the presentation be cut off so that the reader is not overwhelmed with details that may not apply to them?

187

Which environmental management systems are essential for good environmental protection and certification and which are extraneous? This book has attempted to answer these questions and to highlight only the critical systems.

Care was also taken not to concentrate on specific types or sizes of industries, with their own unique impacts and controls. This book was written to apply in general to most organizations and industries. The elements and procedures suggested are generic and would be as appropriate for an oil refinery as for a small printing firm, for example. The difference would be that the organization with significant impacts should develop most of the suggestions to a greater degree than a company with minor impacts.

Even if an organization is not interested in an ISO 14001 certification, the environmental management systems presented in this book should still be implemented. This is due to the fact that the procedures and elements are based upon sound environmental management principles. Implementation will benefit the environment and business in terms of reduced risks and long-term costs. Special attention was directed toward making these environmental procedures practical. Most have been tried by the author and found to be down-to-earth and common sense. If these suggestions are followed, whether for ISO 14000 or not, there will be a significant positive impact on the environment and the organization.

Hopefully the reader has sensed that there is an incredible amount of energy and momentum involved with worldwide ISO 14000 development. The great number of committees alone is a sign of the high level of effort. These committees exist at the international, national and regional levels. In addition to the international subcommittees to TC207, there are as many as 30 different countries with national committees. Many of the subcommittees have over 50 different organizations represented per committee. Some of these organizations have individuals who are working almost full time in the development or tracking of ISO 14000.

This book has been organized in a way to allow the reader to prepare the elements and procedures needed in the most cost-effective manner possible. The background necessary to understand what ISO 14000 is all about was first presented. This included a presentation of the scope, benefits, committees involved and the certification process. Once the background was addressed, some actual components and procedures required for ISO 14001 certification were given. Examples of policies, objectives and targets were provided along with environmental management procedures. These procedures could be used as a starting point by most organizations, who could then add their own site-specific elements.

Since this book has concentrated on only the first component of ISO 14000, that being ISO 14001 or (SC1)—Environmental Management Systems, there are still many important environmental topics yet to cover. Environmental auditing (SC2), environmental labeling (SC3), environmental performance evaluation (SC4) and life-cycle analysis (SC5) are also part of the ISO 14000 plan and are optional at this point in time. SC2–5 are busy developing these standards at the present time and have started issuing drafts. The reader is advised to closely watch the development of these standards since they may be future certification requirements and even if they are not, they provide ways to continually improve the overall environmental management system. These topics have been only briefly mentioned in this book. Once the SC2–SC5 standards are a little further along, it will be time for organizations to start working toward implementation of these elements as well. The proactive organization will start to incorporate as many of these future elements into their systems as early as they can.

This book has provided environmental management procedures which should address the new ISO 14001 standards. An emphasis was placed on environmental management systems (SC1). The book is a step-by-step approach to improve environmental management systems and to meet the certification requirements. The author wishes the reader success in joining with thousands of concerned people and businesses around the world in the challenge to maintain business excellence while protecting the earth and our quality of life.

Appendix A
Sample Job Descriptions: Environmental, Health and Safety

Environmental, Health and Safety Director

- Oversee financial aspects of soil and ground-water cleanup projects
- Interface with public relations department
- Interface and develop strategies with the legal department
- Report key environmental, health and safety indices to senior management
- Set strategy for corporatewide industrial hygiene program
- Ensure new regulations are being interpreted and reported to operating units
- Develop overall direction for major environmental, health and safety programs
- Coordinate the development of standard operating procedures
- Interface with agencies and insurance groups
- Set overall direction for waste minimization program
- Ensure treatment, storage and disposal vendors are audited and are a reasonable risk
- Direct environmental site assessments for new properties
- Manage the review of drawings, plans and designs for new or modified systems
- Oversee workers' compensation program activities

- Report environmental, health and safety problems and progress to senior management

- Determine overall site compliance strategies

- Evaluate site performance to operating procedures and regulations

- Audit sites in terms of environmental, health and safety compliance

Environmental Manager

- Provide direction and assistance to operational individuals

- Assess facility closure activities for compliance

- Review all new property transactions for environmental issues

- Manage soil and ground-water cleanups

- Review regulations and design compliance programs

- Prepare or ensure that all agency permits and reports are submitted on time

- Analyze monitoring data and required report information

- Manage the waste minimization program

- Audit hazardous waste treatment, storage and disposal facilities

- Audit operational sites to ensure compliance with regulations

Environmental Engineer

- Inspect chemical and hazardous waste areas

- Prepare and update hazardous material management plans

- Supervise emission control equipment

- Complete emission reports

- Complete hazardous waste reports

- Prepare and maintain emission licenses and permits

- Audit locations for regulatory compliance
- Conduct environmental training
- Update environmental manuals
- Sample waste streams
- Assist in reducing chemical and resource needs
- Maintain all spill equipment
- Promote waste minimization

Appendix B
Register
of Regulations

REGISTER OF REGULATIONS COVER PAGE

DATE:

REV: 001

DOCUMENT NUMBER:

QUALITY/ENVIRONMENTAL MANAGER:

DESCRIPTION: This binder contains regulation summaries (or the actual regulations) that apply to site operations.

ALTERATIONS: Alterations are not permitted without prior approval of the Environmental Manager and must be applied using the system for amendment control contained within this document.

Verification:

	Signature	Function	Date
Complied By:			
Approved By:			
Authorized By:			

Location of Regulations:

	In Paper File	In Binder	In Computer File
International			
Federal			
State			
Regional			
Local			

REGISTER OF REGULATIONS SAMPLE INDEX

1. Procedure to Identify, Analyze, Address and Record Regulations

2. International Regulations

    2.1. Montreal Protocol

    2.2. Basel Convention on Control of Transboundary Movement of Hazardous Waste and their Disposal

3. Europe Regulations

    3.1 British Standard 7750

    3.2 COSHA

    3.3 Clean Air Act of Northern Ireland

    3.4 Others

4. United States Regulations

    4.1 Clean Air Act

    4.2 Clean Water Act

    4.3 Resource Conservation and Recovery Act

    4.4 Toxic Substances and Control Act

    4.5 Pollution Prevention Act

    4.6 Others

5. State Regulations

6. Regional Regulations

7. Local Regulations

Appendix C
Treatment, Storage and Disposal Facility Audit Form

TREATMENT, STORAGE AND DISPOSAL FACILITY AUDIT FORM

Date of Inspection _____

Name of Inspector _____

General Information

Vendor Name _____

Date of Inspection _____

Address _____

Contact Name and Telephone _____

EPA ID Number _____

Name of Inspector _____

Description of Operations (Including Materials Handled)

_____

_____

_____

_____

_____

_____

_____

Facility Owner _____

Site Manager _____

Parent Company Name _____

Parent Company Address_____

History of Facility _____

_____

_____

_____

Major Customers_____

_____

_____

_____

General Land Use within Five Miles _____

_____

_____

_____

Proximity to:

       Hospitals _____

       Endangered Species _____

       Schools _____

       Floodplains _____

       Waterways _____

       Residences_____

Insurance in Place (Type and Amount) _____

_____

_____

Financial Stability (D&B) _____

_____

Qualifications of:

    Managing Director _____

    Lab Manager _____

Regulatory Information

Agencies with Authority _____

_____

_____

	Agency	Expiration	Conditions

Current Permits

    RCRA Part B _____

    Air _____

    Water _____

    Transportation _____

    Other _____

Maximum Storage Allowed _____

_____

_____

Closure Plan Present _____

_____

_____

Agency Inspections _____

_____

_____

Complaints Filed _____

_____

_____

Enforcement Actions Filed _____

_____

_____

Fines _____

_____

_____

Lawsuits _____

_____

_____

Cleanups (Ongoing or Planned) _____

_____

_____

Manifests on File _____

_____

_____

Required Training Completed _____

_____

_____

Written Training Program and Records on File _____

_____

_____

_____

Facility Information

Warning Signs _____

_____

_____

Security Procedures and Equipment _____

_____

_____

Fences and Natural Barriers _____

_____

_____

On-site Buildings and Structures (Condition and Use) _____

_____

_____

_____

_____

Fire Prevention Systems (Type, Capability) _____

_____

_____

_____

Written Fire Prevention Plan _____

_____

_____

Personal Protective Equipment Available (Type and Condition)

_____

_____

Odor Control _____

_____

_____

Ground-Water Monitoring Wells

    Number _____

    Type _____

    Location _____

    Depth _____

    Testing Frequency _____

    Average Results _____

    Standards Measured _____

Container Storage Area

    General Condition _____

    Secondary Containment _____

    Evidence of Leaks or Discharges _____

    Inspection Records _____

    Condition of Containers _____

    Segregation of Incompatible Wastes_____

    Overall Protection of the Area _____

    Volume Stored (and Allowed Amount) _____

    Covers _____

	Aboveground	Underground

Tanks

General Condition _____

Secondary Containment _____

Leaks or Discharges _____

Volume Stored
    and Allowed _____

Inspection Records _____

Cutoffs _____

Leak Detection Systems _____

Monitoring Systems _____

Tank Testing Records _____

Laboratory Capabilities

Quality Control _____

On-Site Analytical Capability _____

Off-Site Labs Used _____

Major Instruments On-Site _____

Waste Acceptance Tests Done _____

Waste Acceptance Documentation Needed _____

Incineration Capabilities

Type of Incinerator _____

Wastes Handled _____

Scrubber _____

Ash Disposal Location _____

Permits for Air Discharge and Bottoms _____

Burning Prohibitions _____

Inventory Control _____

Disposal Areas

    Liners _____

    Leachate Collection Systems _____

    Storm Runoff Collections Systems _____

    Ground-Water Monitoring Wells _____

    Segregation of Wastes _____

    Inventory of Wastes _____

Other Types of Treatment Systems

    Description of Operation _____

    Permits _____

    Emission Controls _____

    Residue Handling and Disposal _____

    Drum Inventory _____

Appendix D
Assessment Procedures

Introduction

The term "assessment" is being used in this chapter to refer to general investigative activities that are done to better understand the impacts of HMW. Sampling and analysis, which were just discussed, are some of the first steps in the assessment process. The property transfer assessment, operational audit and risk assessment all help to define the impacts of HMW on current operations. This chapter will concentrate on these three assessment types.

There are numerous types of environmental reports and assessments that traditionally have not concentrated on HMW. A few examples were discussed in Chapter 1. These included the environmental assessment (EA), the environmental impact report (EIR) and the environmental impact statement (EIS). Usually, these documents are required prior to construction of major projects (and regulatory changes) and deal with general environmental and socioeconomic impacts. Even if the EA, EIR or EIS are not discussed in this chapter, they are shown in Table 7–1 so that the reader can see the relationship between them and the assessments that will be discussed (i.e., the property transfer assessment—PTA, operational audit and risk assessment). Table 7–1 arranges the different assessment types from top to bottom in terms of timing (year 1 to year 20 or 30 of an operation) and from left to right in terms of severity of problems noted.

Reprinted from *Practical Management of Chemicals and Hazardous Wastes,* pp. 106–128, by Lee Kuhre. © 1995 Prentice Hall PTR, Upper Saddle River, N.J., 07458.

Table 7–1 Comparison of Several Common Types of Assessment

Timing	Name of Review	If Problem Noted
Prior to purchase	Phase I Property Transfer Assessment (PTA)[a]	Phase II & III PTA[a]
Prior to construction	Environmental Assessment[b]	EIR and/or EIS[b]
After operation for a period of time	Operational audit[a]	Risk assessment[a]
Prior to sale	Phase I PTA[a]	Phase II & III PTA[a]

[a] Emphasis of the assessment is hazardous materials or wastes.
[b] Emphasis of the assessment is project impacts.

Property Transfer Assessments

A property transfer assessment, due diligence review, prepurchase or pre-lease review are all similar terms. This type of review should be done for any commercial, industrial or multiple-family development prior to the purchase, lease, financing or sale of the land or building. This will minimize later impacts to the environment and reduce the risk of potential litigation, especially concerning hazardous waste or contamination left by a previous owner.

Environmental Liabilities in Property Transfer

Environmental liabilities during property transfer arise from various statutes. Of greatest importance is CERCLA, or Superfund, which establishes liability that is retroactive, eternal and can't be contracted away. In addition, SARA, RCRA, Clean Water Act, the Clean Air Act and other acts involve property transfer liabilities.

The level of liability that is involved in property transfer can be astronomical. The PTA will help minimize that liability. Some organizations that have not done a PTA have found themselves the owners of ground water cleanups costing many millions of dollars. Injuries and law suits can also be unwelcome aspects that accompany the deed to a property.

Due Diligence

Due diligence is a flexible use of processes and techniques intended to allow an interested party to evaluate the potential environmental risks associated with business transactions involving real estate. It is a risk management tool. It uncovers the likelihood of hazardous substance contamination on the property, the potential environmental liabilities associated with conducting the business or owning the property and the value of the property.

Due diligence developed primarily in response to CERCLA or Superfund. CERCLA states that parties who have had no involvement with the hazardous substance contamination of property may be found liable for very expensive cleanups merely because they own the property.

The extent of the due diligence survey is important. The survey must be consistent with good commercial or customary practice. It must be taken a step further when a reasonable person would be suspicious of environmental problems. The American Society of Testing and Materials has introduced standards for property assessments (E1527-93 and E1528-93). ASTM standards are often used in court, though there is no guarantee this standard will be adopted universally in terms of survey extent.

Innocent Landowner Defense

The innocent landowner defense is specified in the Superfund Amendment and Reauthorization Act (SARA) and is the only partial relief to the CERCLA liability problem. To qualify for the defense, the landowner must not have known that the property was contaminated at the time of acquisition. They must have made reasonable inquiries into the past uses of the property before acquisition to determine whether the property was contaminated (environmental due diligence). The property must have been acquired by the defendant after the disposal or placement of the hazardous substance. They must have reacted responsibly when the contamination was found. As a practical matter, qualifying for the defense is very unlikely under most circumstances.

In addition to the innocent landowner defense, there is an exemption for parties that hold title to property because they have a security interest in the property. The EPA has published a regulation indicating what actions these parties can take without losing their protected status (40 CFR 300).

Levels of a Property Transfer Assessment

There are varying degrees of depth at which a property transfer assessment can be done. These range from a very general overview to an in-depth study. The different levels are shown in Table 7–2 and include an initial screening and Phases I, II and III. In general, an initial screening does very little; a Phase I identifies potential problems; a Phase II defines the problems; and a Phase III provides costs and correction. Phases I, II and III are defined very loosely, so it is important that the consultant spells out very clearly what will be covered in each phase.

Table 7–2   Levels of Assessment

- Initial Screening—Quick visual overview
- Phase I (Tier I)—Identification of potential problems
- Phase II—Characterization of identified problems
- Phase III—Detailed cost analysis (and corrections)

Public Records Review

Table 7–3 presents some of the records that should be reviewed during the record review portion of a Phase I property transfer assessment. Of all the records mentioned, the federal (see II.A.1) and state (see III.F.1) Superfund lists and regional agency lists of leaking tanks (see IV.B) are the most important. Aerial photographs and Sanborne Fire Insurance Maps are the most informative for historical information. Old city directories that are arranged by street address are also very useful, but have only been prepared for certain locations and years.

Aerial photographs were one example of a source of the historical information just mentioned. Figure 7–1 is an aerial photograph taken on July 23, 1980. By studying the original photo, especially with magnifica-

Table 7–3   Public Records Review: Due Diligence Checklist

I. Real Estate Records
    A. Title Reviews
        1. Grantor/grantee 50 year chains (or longer if records are available)

II. Federal Sources of Information
    A. EPA (and in some cases state)
        1. Superfund National Priorities List—NFL Sites
        2. Comp. Emergency Response, Compensation and Liability Information System—NFL sites and sites which may be contaminated
        3. CERCLA Release Notifications and Follow-up Reports
        4. Emergency Remedial Response Information Center System—sites potentially contaminated
        5. Toxic substances Release Inventory—releases of a toxic chemical
        6. TSCA Inspection and Compliance Records—PCB violations
        7. Violator List—Violations of various environmental laws
        8. NPDES and POTW approved discharges
        9. Petroleum and Oil Spill Reports
        10. UST Records
        11. EPA ID number—generators
        12. Asbestos notifications about demolition, renovations, and releases
    B. U.S. Army Corps of Engineers
        1. Section 10 permits
        2. Section 404 permits
    C. USGS
        1. Ground water information
    D. Security and Exchange Commission
        1. Filings regarding potential liabilities
    E. Emergency Planning and Community Right to Know Agencies
      (Fire Department, EPA, State)
        1. Stored hazardous chemicals and extremely hazardous waste

III. State Sources of Information
    A. Air Quality
        1. PSD permits
        2. Air construction permits
    B. Water Quality
        1. NPDES permits
        2. Enforcement actions
    C. Coastal Zone Management
        1. Building permits
        2. Wetland issues

Table 7–3   Public Records Review: Due Diligence Checklist (Continued)

    D. Natural Resources
        1. Oil, gas, and mineral development permits
    E. Solid Waste Management
        1. Disposal of nonhazardous solid waste
    F. Hazardous Waste Management
        1. TSDF permits and Superfund list
        2. Audits and Enforcement violations
    G. Fish and Wildlife
        1. Damage records
IV. Local Sources of Information
    A. County, City, Fire Departments
        1. Building permits
        2. Land use permits
        3. UST —HMMP
    B. Regional Water Quality Control Boards
V. Other Sources of Information
    A. Aerial Photographs
    B. Computer Data Firms

tion and in stereo, the industrial operations, waste disposal areas, above-ground storage tanks and other surface features can be noted. Sometimes, it is even possible to distinguish stained soil, stressed vegetation and unauthorized drum-dumping areas.

Field/Site Review

Table 7-4 lists some of the things to look for during the field or site review portion of a Phase I [Denton 1989]. The field inspector must be trained to look for subtle clues of present and past contamination. Most HMW real estate problems of importance are not going to be obvious.One item not shown in Table 7-4 that would be important to check is the drawings for the operation going all the way back to the first-year grading/earthwork drawings. This may show that, a few years ago, a sump, tank, trench or other structure was installed. It may now

Figure 7–1 Aerial photograph. (Pacific Aerial Surveys, 1980. 8407 Edgewater Dr., Oakland, California 94621)

be out of service and covered. It also may have leaked when it was being used. Usually, there would be no way of seeing either the structure or leak by walking the site. This makes a review of the past drawings very important.

Table 7–4   Site/Field Review: Due Diligence Check List

I. UST and Piping—Check HMMP and building permits for:
    A. Age
    B. Type—Problems with bare steel tank and piping
    C. Spills and Overflows

II. Asbestos—Check for:
    A. Demolitions and renovations
    B. Releases
    C. Presence of ACM
    D. Age of building

III. PCBs
    A. Age of electrical equipment
    B. Electrical work areas
    C. Electrical disposal areas

IV. CERCLA Hazardous Substances
    A. Any amount of the 717 substances listed in 40CFR Table 302.4

V. Processes
    A. Industrial
        1. Type of past and present processes
        2. Permits adequate, current, acceptable limitations and constraints
        3. Pollution control equipment o.k.
        4. Grandfathering of equipment
    B. Waste Management and Disposal—Compliance of past and present processes and waste streams
        1. Treatment and disposal into POTW
        2. Treatment and disposal via NPDES
        3. Disposal via Class I, II, or III
        4. Recycling or Reuse
        5. Has waste minimization been practiced?
        6. Are environmental systems at capacity or is there some growth potential?
        7. What kind of management support exists?
    C. Obligations for Mitigation Efforts

VI. Wetlands and Flood Plains
    A. Proximity to Water
    B. Frequently Inundated with Water
    C. Army Corps of Engineers Permits
    D. Fill Activities in the Past
    E. Dredge Activities in the Past

Table 7–4   Site/Field Review: Due Diligence Check List (Continued)

VII. Hazard Communication Program Records A. List of hazardous chemicals B. MSDS C. Proper handling, warning, emergency planning VIII. Radon A. Short term test or "grab sample" (immediate)

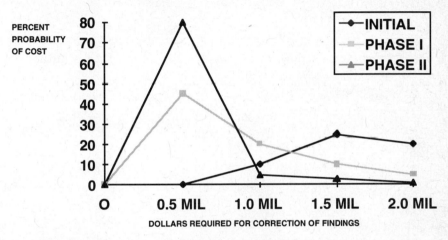

Figure 7–2 Property Transfer Assessment: Cost to Correct Findings.

Types of Facilities Requiring Review

A Phase I should be done prior to any real estate transaction involving all types of commercial, agricultural, industrial and multiple-family residential properties. The depth depends on the type of operation. At the present time, a Phase I is not strictly required, just inferred or recommended. However, most real estate transactions require the seller to fill out a property disclosure checklist, listing knowledge of environmental problems.

Costs Identified During a Property Transfer Assessment

Figure 7–2 summarizes all phases of a property transfer assessment in terms of cleanup cost for a site in Illinois. As can be seen, the initial screening does not provide much information, especially in terms of cleanup cost, and is a more conservative estimate. The Phase I assessment starts to show some clarity in cleanup cost (range from $.5 million to $1million). In this case, the Phase II assessment nailed down the cleanup cost at about $500,000. Since the curves never meet the x-axis, it must be understood that there is always a chance that the costs could go lower or higher.

Operational Audit

In the preceding section, we were at a pre-lease, preloan, prepurchase point in time. In this section, the organization is further along in time and has already purchased or leased and has operated at the site for a period of time. It is now time to analyze or audit the operation in terms of ongoing impacts and compliance. This includes the organization's own operations and contractor operations.

From a regulatory standpoint, it is as critical to do a routine compliance check on operations as it is for an individual to have a routine physical exam. With laws and regulations changing almost daily, an operation can quickly move into noncompliance. In addition, the U.S. EPA has a formal policy encouraging the development, implementation and upgrade of auditing programs. The EPA will take into account the existence of such programs when determining the appropriate penalty in the event of a violation. In certain EC countries in Europe, the eco-audit regulations require that operations perform audits as of January 1, 1994.

The suggestions presented in this section could apply to contractors as well as an organization's own operations. There should be a good agreement in place with the contractor that protects the organization. The agreement should hold the contractor responsible for compliance with all environmental, health and safety laws and regulations. In addition, the contractor should have considerable environmental insurance or adequate self-insurance capability. The agreement does not completely remove the necessity to do occasional contractor audits. The audits should not violate the normal "independent contractor" condition in most agreements.

An operational site analysis or audit is a systematic assessment of compliance. It is the process of determining whether all or selected levels of the organization are in compliance with the regulations, internal policies and accepted practices. It is a check of the regulatory status of an individual facility.

Levels or Degrees of Depth

Different organizations invest varying degrees of time in operational audits. This is dependent on issues such as past fines, available resources, time and number of hazardous waste streams. The following four levels of audit are recommended, in decreasing order of importance:

- *Known contaminant impact analysis*—If a problem is known or suspected an in-depth analysis should occur to determine the quantity and extent type of contamination.
- *Regulatory compliance analysis*—A regulatory compliance analysis should be done by all organizations. This involves an assessment of compliance with all HMW laws and regulations in reference to a specific site.
- *Corporate policy compliance*—An assessment of corporate compliance should be done by most organizations, if possible, assuming that they have established HMW policies over and above regulatory requirements.
- *Best management practice compliance*—Best management practice for the industry should be compared to an organization's practices if significant hazardous waste streams are present. This is done in order to stay ahead of the regulations. It reflects generally what the rest of the industry is doing about specific hazardous wastes, especially in terms of treatment. Best Available Control Technology (BACT) is one way of describing treatment levels.

Key Principles

Many principles should be considered when designing an operational audit system. For example, what is the objective, and is it achievable? Will the system be systematic and supported by evidence? Results must be reported understandably, quality assurance/quality control (QA/QC) followed, and the field individuals must be involved in the process. The analysis must be fair, measurable and done in a nonthreatening way. It should not contain opinions or judgments. The analysis should be done at a set frequency, with some surprise visits as well.

It is important to consider the legal ramifications of audits. For example, if problems are uncovered and documented, they must be promptly corrected. If they are not, these "smoking guns" can do an organization great damage. In other words, a documented problem that is not corrected is a prime candidate for a fine or court action.

Total environmental quality management (TEQM) audits are an attempt to bring total quality management principles to bear on environmental problems. TEQM was pioneered by the Global Environmental Management Institute. In TEQM audits, the primary focus is on systems rather than specific compliance. For example, in the traditional audit, the inspectors review all HW drums for proper labelling and then document any deficiencies. In the TEQM audit, they take the additional step of asking whether systems are in place to ensure that HW drums are labelled 100% of the time. With this emphasis, the people being audited are less likely to look at the audit team as nitpickers from the home office and are, in fact, an important part of the audit team.

Audit Team Composition

There should be careful thought concerning who should audit, since different situations dictate different types of assessors or auditors. First, the auditors should be qualified and have a good understanding of internal processes and controls. If one does not understand the internal controls, it is extremely hard to verify that they are functioning. It may be appropriate for the auditors to be from an agency or a consultant, company staff, or field organization. They could be either part-time or full-time. However, it is hard for the part-time auditor to stay on top of all the applicable laws and regulations. Each situation usually dictates a different type of auditor or audit team. In the author's experience, an internal audit team should be composed of at least one regulatory expert and one field representative who knows the site and the process.

In some parts of the world (Europe, for example), there is proposed legislation that would require that an independent auditor assess for operational compliance with hazardous waste regulations. If the assessment is done, corrections made and results offered to the public, then the operation can obtain a "green" status or bill of health.

On an even broader scale, the ISO quality certification is starting to include environmental policies and procedures. Similar to the above example, an independent auditor must assess a company operation; if it meets quality criteria, it would obtain a certification. This certification and the above "green" status are very important for staying in business in Europe and are starting to become an issue in the U.S. as well.

Common Analysis/Audit Tools

There are some tools that are used during most audits. Usually, an auditor will select one or two that best fit the situation. Some of the choices include a topical outline (lists topics to be covered), detailed guide (lists regulatory requirements plus standards), yes/no questionnaire, open-ended questionnaire (explanation required) or scored questionnaire (responses are scored against criteria). No matter which tool is selected, it is important that the auditors use the tool. Far too often, auditors will "wing it" or just walk the site and expect their memories or experience to remind them what to look for. This is a risky habit to get into, since there are hundreds of things to consider.

Table 7–5 lists some possible HMW issues that would be assessed during a typical audit. In addition to the suggestions listed, state and local requirements should also be included. This list is not meant to be all-inclusive, just to provide some examples. Actual audit lists that the author has used in the past range from two pages for an administrative office up to 20 pages for a large manufacturing facility. Most audit lists would also include columns for suggested corrective action, scheduled completion date, actual completion date and responsible individual.

Table 7–5 Operational Audit: Examples of Hazardous Material
and Waste Issues to Assess

Air Pacs	Air pacs (SCBA) are full, inspected monthly and tagged
Alarms	Evacuation alarms are present
Audits	Hazardous waste treatment workers audit & record daily
Audits	Supervisors complete and document general safety and hazardous waste storage/treatment area weekly
Chemical Labels	All chemical containers/vessels are labeled with hazardous contents and appropriate warnings
Chemical Exposures	Medical surveillance and IH data is collected for employees exposed to anything that may approach PEL
Chemical Handling	Chemical handling is centralized
Chemical Inventories	SARA Title III inventories are completed and submitted on schedule each year
Chemical Spill Audits	Weekly chemical audits are completed and documented
Chemical Storage Secondary Containment	Chemical storage areas have secondary containment or absorbents
Chemical Storage Segregation	Incompatible chemicals aren't stored together
Chemical Storage Room	Room or area is well ventilated and clearly posted with signs
Chemical Transporting	Chemicals in glass containers are transported on-site via carts or rubber carriers
Chemical Volumes	Quantities stored comply with UFC. No more than 1 day supply of flammables or corrosives outside of the main storage area
Complaints	A procedure is in place to handle HMW complaints
Compressed Gas	All cylinders are capped and secured
Concealed Hazards	Significant concealed hazards are disclosed to affected employees immediately. Faulty system removed from service, repaired or OSHA notified in 15 days
Containers	No open containers of chemicals or waste
Discharge	No discharge to streams, drains or sewers unless analysis on file that shows compliance
Disciplinary Action	Applied for violation of laws and procedures
Dust Collection	Respirable dusts are vented to a collection system
Emergency Action Plan	A written plan is in place to address spills, tank ruptures, etc.

Table 7–5 Operational Audit: Examples of Hazardous Material
and Waste Issues to Assess (Continued)

Emergency Coordinator	An on-site person has been designated
Emergency Notification Numbers	Numbers are posted throughout and in all required documents (manifests, HMBP, etc.)
Emissions Reporting	All required discharge data is reported
Empty Chemical Containers	Residual chemicals are used and container is washed or returned to supplier
EPA ID Number	On file for the site and used during manifesting
ERT Drills	ERT performs monthly emergency drills
ERT Plans	Plan contains current names, numbers, etc.
Evacuation	Adequate, up-to-date evacuation maps are posted
Eye Protection	Eye protection is available where chemicals and wastes are handled
Eyewash/Shower	Functioning unit is within 25' of chemical handling or storage areas
Fire Extinguishers	Extinguishers are within 10 feet of flammable storage
First Aid Assistance	A nurse or EMT is available on-site
Hazardous Waste Storage Time	Waste is not stored beyond the 90 day, 180 day or 365 day period specified
Hazardous Contingency Plan	A management or business plan is on file which deals with chemicals, wastes and emergencies
Hazardous Waste Storage Communication	Internal and external communication is provided in storage and treatment areas
Manifests/Records	All hazardous waste has a properly prepared manifest and LDR forms. Exception reports, biennial reports and waste determinations are kept on file for at least three years
Material Safety Data Sheets	MSDSs are available at the location where chemicals are present and employees are aware
Notifications	Employees are notified of chemical exposure levels. Medical files are up-to-date
Permits/TSDF	A TSDF permit or permit-by-rule is on file if hazardous waste is stored over 90 days or treated on-site
Permits/Water Discharge	All necessary permits for discharge to streams, storm run-off and sewers have been obtained
Personal Protective Equipment	Surveys, fit testing, training and follow-up done to insure proper use of PPE

Table 7–5   Operational Audit: Examples of Hazardous Material
and Waste Issues to Assess (Continued)

Pumps and Pipelines	No leaks are present and they are checked on a routine basis
Respiratory Protection Program	Appropriate respirators are present and maintained after surveys, fit testing and training
Risk Assessment	An assessment of risk has been performed where there is potential significant risk
Signs	Required signs are posted, such as Hazardous Waste, etc.
Spill Clean-up Equipment	Absorbents, neutralizers, gloves, etc. are present where fuel, chemical and waste are used or stored
Sumps	Sumps and secondary containment berms are sealed, lined, free of liquids and inspected routinely for cracks and so forth
Training	Employees are properly trained and tested in chemical handling, hazard communication, first aid, etc.
Vendors	Hazardous waste vendors are registered and checked-out prior to use and at routine intervals thereafter
Ventilation	Adequate exhaust ventilation exists where needed
Waste Characterization	Wastes have been tested and there is an analysis on file regarding whether waste is hazardous
Waste Minimization	A pollution prevention program is in place

Audit Process Steps

The actual audit process steps range from very simple and straight-forward for small sites without many waste streams to very involved procedures for large facilities. The following generic checklist of activities would apply, in general, to both ends of the spectrum. The amount of time spent in each phase could range from several hours to several days.

Pre-Audit Activities

There are some actions that should be done before visiting the site, by the audit team. These include selecting the audit team, selecting facilities (mini-risk assessment), scheduling the audit and gather background information for the team (paper search). Most important, it is necessary to contact the site management before the audit. Their schedule, sensitivities and needs must be considered for the audit to be successful. It may even be advisable to not refer to it as an audit, since this word holds a negative connotation for many people. It could be termed an "operational review" or "compliance check."

Field Analysis/Audit

Once the upfront activities are done, the field work can start. The following actions are done at the site by the audit team and site individuals. The first action is usually a kick-off meeting at the site. This lets everyone know the purpose and schedule for the day. Next, interviews and other on-site information gathering occurs. This is followed by the most important step, the location walk-through. Audit findings are entered into a checklist during the walk-through. The final step at the site is an exit meeting to review and summarize the initial findings.

Post-Audit, Followup and Correction

Post-audit actions are done after the field work by the audit team and site individuals, with as much involvement of the latter as possible. The audit team will first analyze and interpret the findings. The checklist or action plan is usually prepared to commit to specific actions by specified times. If they are big problems, a feasibility study and remedial action plan may be necessary. The audit team should then distribute the checklist or action plan. Far too often, it all stops here. Uncorrected findings (smoking guns) can lead to real liability problems. Correction and followup to ensure that the problems and their root causes are fixed are of paramount importance. An example of a followup checklist form can be found in Table 7-6.

Table 7–6   Audit Action Item Followup Checklist

Problem	Suggested Correction	Proposed Completion Date	Actual Completion Date	Corrected by Whom

It is important that even if a long report or action plan is prepared, a checklist is also completed. OSHA has promoted the use of checklists. It is much easier for the site individuals to fix a list of items, each with their own correction action, date and person assigned. It is also easier for the auditor to track this type of checklist to ensure completion. Auditors should not prepare long paragraphs full of various actions, all of which are to be completed by unknown individuals on one unrealistic date. It is an open invitation for failure.

Examples of Common Problems Noted in Audits

The types of problems uncovered in audits are numerous and could fill several volumes. The author has noted the following trends after working in a variety of industries: lack of training or understanding, lack of control and tracking systems, inadequate facilities, lack of management systems, higher than acceptable employee or public exposure, procedures inadequate or not understood, environmental pollution occurring, hazardous material source reduction not practiced, improper storage of hazardous materials and missing permits, plans and variances. Record-keeping deficiencies are a common problem, along with lack of resources (money and staff).

Ongoing Monitoring

An operational audit, a permit requirement or various regulations may require an organization to do ongoing monitoring. This may involve ground water, process discharge, worker exposure, surface water and many other types of monitoring. Only by monitoring can an organization really know where they stand in terms of protecting the environment and workers' health. This also allows the organization to make corrections early, before problems get out of hand. Please refer to Chapter 6, "Sampling and Analysis," for details that apply to monitoring.

The only self-monitoring that will be discussed further is for ground-water monitoring. All TSDF permit holders are required to have sufficient wells to assess the quality of background water and detect any possible contamination from the operation. Indicator parameters, waste constituents and reaction products must all be measured.

Risk Assessment

Many organizations are starting to do risk assessments in reference to some of their HMW. Certain regulations, such as California Prop. 65, require a warning if there is significant risk of exposure. The risk assessment allows this exposure determination to be made. Indirectly, risk assessments are also required by CERCLA, SARA, the National Contingency Plan and other regulations.

The discussions in this section concentrate on health risk assessments. There is, however, a whole new field developing: ecological risk assessments. As the names imply, health risk assessments estimate the impact of a chemical(s) on a human population, while ecological risk assessments estimate the impact on the entire ecosystem. The two types are compared in Table 7–7.

Table 7–7   Risk Assessment Comparison

Health Risk Assessment	Ecological Risk Assessment
■ Hazard Identification	■ Receptor Characterization (species, life stages, etc.)
■ Dose-Response	■ Hazard Assessment (nature and toxicity of effect)
■ Exposure Assessment	■ Exposure Assessment
■ Risk Characterization	■ Risk Characterization

Risk assessment helps to estimate the likelihood of injury, disease or death resulting from exposure to a potential hazard. Since there are many unknowns and uncertainties in a risk assessment, there is a need for more uniformity and standardization. The four main elements of a health risk assessment are illustrated in Figure 7–3.

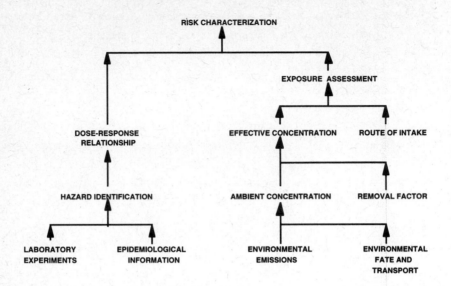

Figure 7–3 Risk Assessment.

Hazard Identification

In this stage of risk assessment, one would review, analyze and weigh toxicity and carcinogenic effect data in various settings. Just as the title implies, the goal of this first step is to identify the hazard(s) or base cause of the problem. Some researchers refer to this first step as the toxicological evaluation. It is a qualitative evaluation to determine whether the chemical may have a potential adverse effect.

Types of Studies

Human, animal and cell culture studies are the sources of the raw data. Human studies utilize case reports and epidemiological studies. The epidemiological studies are better but still not controlled, since they are usually from an accidental exposure.

In animal studies, mice, rats, rabbits and other animals are used and then one must extrapolate to humans. Sometimes the correlation is low between species. Animal studies include acute studies (short-term, high-level exposure) and chronic dose studies (life-time, low-level exposure).

Cell culture studies utilize test tubes or flasks containing cells. The cells are usually attached as monolayers or free-floating. For example, it is possible to get monolayers of skin and liver cells to grow in cultures. The HMW is then introduced onto a specific number of growing cells. Extrapolation has to be to the whole organism, which is hard to do since there is no organ interaction. One advantage of cell culture studies, however, is that impacts on cellular functions can be seen fairly quickly. Another advantage is that it reduces the amount of animal studies necessary along with the associated animal suffering.

Types of Effects to Analyze

During the above-mentioned studies, many observations are made. The most important categories of effects are usually mutagenesis, carcinogenesis, teratogenesis, weight loss and liver damage. Irritation, asphyxiation and sensitization are also effects that are commonly assessed.

The carcinogenic effect is usually the driving factor in risk assessment, because many people feel there is no threshold dose (as little as one molecule may cause cancer). In addition, a high degree of uncertainty requires low acceptable or worst-case limits. One way to classify carcinogens is by the weight of evidence system, which includes known, probable, possible, questionable and negative evidence.

The Dose-Response Relationship or Evaluation

The next phase of risk assessment is the dose-response evaluation. At what dose is the substance toxic or carcinogenic? The purpose of this phase is to estimate the incidence of adverse effect as related to the magnitude of human exposure to a substance.

The most important part of a dose-response evaluation or curve is the lower dose area, an area where few if any data points exist. To address this problem, many mathematical models have been developed. The mathematical models are used for extrapolating data to the important lower doses. All of the carcinogen curve models illustrated in Figure 7–4 are non-EPA-approved except the Linear Multistage Model [Wentz 1989].

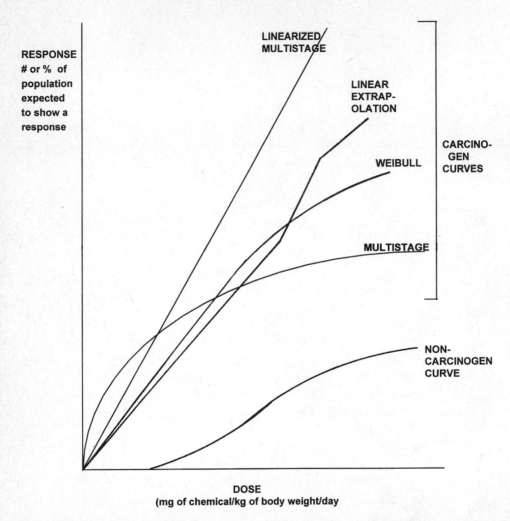

Figure 7–4 Dose-Response Curves.

The Linear Multistage Model is approved by EPA because it is the most conservative. This model assumes that there is no threshold or reference dose and that any amount (even one molecule) will cause cancer.

The noncarcinogen curve is different from carcinogen curves in both shape and where it crosses the *x*-axis. A reproductive toxicant, for example, would have a dose below which there would be no measurable response. This is called the no observable adverse effect level (NOAEL).

Exposure Assessment

In this phase, one would ask the question how much human exposure may be expected? Agencies such as the EPA are especially open to negotiation at this phase of risk assessment, since there is considerable room for interpretation concerning the chance that humans will be exposed.

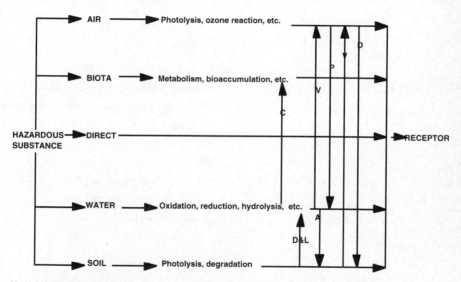

V = Volatization, C = Consumption, A = Adsorption, P= Precipitation, D = Desorption, L = Leaching/runoff

Figure 7–5 Exposure Assessment/Transport and Fate.

First, the source(s) and potential contaminant migration pathways should be analyzed. This helps determine where and how the chemical goes after its release from the source or the environmental fate while it is traveling to the target receptor. The target receptor may be a human, bird, plant and so on. The chemical may be transported in air as a vapor, dust or gas. It may also be transported in water (surface, groundwater, run-off). Transport in plants and animals also occurs, with bioaccumulation in some cases. Last, transport in soil is also a migration pathway. These pathways are summarized in Figure 7–5. As the figure illustrates, a contaminant may move between the major pathways. For example, the process of volatization may take a volatile organic compound from the soil pathway to the air route.

Figure 7–6 "Exposure Pathway Related to Ground Water," is a more in-depth illustration of just part of one line, the water pathway of the overall transport and fate illustration presented earlier. Depending on the contaminant, some of the other components should be broken down as was done for ground water. With this amount of depth and detail, computer programs have been designed to model a contaminant through all the exposure pathways and subpathways.

Figure 7–6 has even missed a few subpathways. For example, during well drilling it is possible that a well will go through two or more independent aquifers. If a well is drilled, constructed or abandoned improperly, it can create a subpathway where contamination is spread between previously independent aquifers. Although it does occur, cross-contamination is limited because of current well construction standards and the awareness of well drillers and geologists of the potential implications. When it does happen, however, transfer of contaminants in this exposure pathway can occur during all types of well drilling, including those shown in Figure 7–6 for domestic, agricultural, industrial, construction and one type not shown, ground water sample wells!

It is important to identify the human exposure points. These include inhalation in air, shower mist and cooking mist. Dermal contact (including injection) is the second type of exposure point. Ingestion in food, water and soil is the last type of exposure point.

An estimate of exposure level or exposure point concentration is the last part of this phase. This can be done by either monitoring and/or modeling. The estimate must consider the extent and frequency of exposure, or how much and how often (short-term or chronic exposure). The amount of absorption, excretion, metabolism and other physiochemical factors should also be factored in. Other variables to consider include the

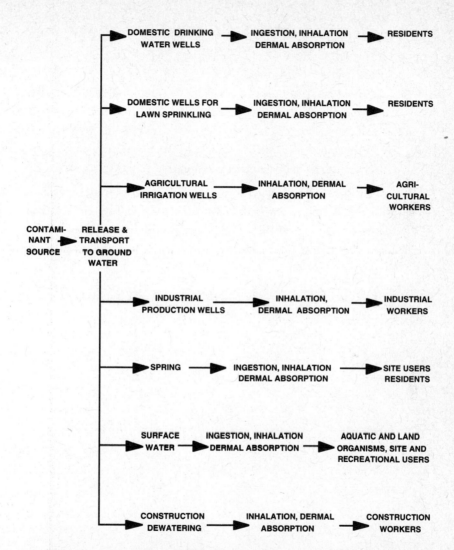

Figure 7–6 Exposure Pathways Related to Ground Water. (From EPA, *Guidance on Remedial Actions for Contaminated Ground Water*, EPA/540/G88/003, Washington, 1988.)

number of people exposed, variation of exposure, degree of intake by various routes, health and age of humans considered, background levels and interaction with other pollutants.

Risk Characterization

The last phase of risk assessment asks, what is the individual or the cumulative population risk? What is the likelihood of injury, disease or death resulting from human exposure to a potential hazard under the conditions assessed during the first three steps of the risk assessment? This is usually expressed as noncarcinogenic or carcinogenic risk. Risk characterization can be calculated for one exposure or cumulative exposures for either an individual or an entire exposed population.

There are several good references on risk assessments. The following are all EPA documents: Risk Assessment Guidance for Superfund Sites, Review of Ecological Risk Assessment Methods, Ecological Assessment of Hazardous Waste Sites and Guidance for Ecological Risk Assessment.

Appendix E
Waste Minimization Procedure

Introduction

There are few topics in HMW management on which all the players agree. With the great variety in background, perspective and purpose, this is not surprising. The one and possibly only subject on which everyone, from the most hardened industrialist to the most radical environmentalist, seems to reach consensus is waste minimization. Waste minimization is the reduction, to the maximum extent possible, of hazardous waste that is generated, treated, stored or sent for disposal.

The EPA has put waste minimization as a top priority in terms of hazardous waste management. Figure 14–1 shows waste minimization as the most preferred option. Resource recovery is shown second and really is part of waste minimization, as is treatment, but to a lesser degree.

From a global perspective, waste minimization is a common theme, and in many respects the Far East and Europe lead the United States. This may in part be due to a shortage of space and resources in those areas of the world. Therefore, there has been a long-standing practical reason to minimize waste. Without question, the Far East and Europe lead the world in nonhazardous waste minimization (such as the German packaging and product take-back regulations) and in many aspects of hazardous waste minimization as well.

Reprinted from *Practical Management of Chemicals and Hazardous Wastes*, pp. 216–228, by Lee Kuhre. © 1995 Prentice Hall PTR, Upper Saddle River, N.J., 07458

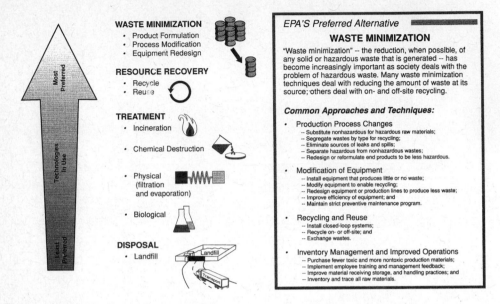

Figure 14–1 Hazardous Waste Management: What's the Best Approach? (Reprinted from EPA, 1993.)

Waste minimization is known by various names, such as waste prevention/avoidance/reduction, pollution prevention and other terms. For the purposes of this book these terms are considered similar, even though there are some differences.

Waste minimization, whether it be for industry, households or any other type of generator, is the way of the future. Land bans and increasing treatment and disposal costs make waste minimization of paramount importance. This chapter will discuss the benefits of waste minimization, priorities, key elements, ranking, options and assistance. The emphasis is on HMW; however, most of the concepts could also apply to nonhazardous material waste minimization.

Benefits of Waste Minimization

Reduced Environmental Impact

Waste minimization results in reduced environmental impact since it reduces waste in the environment. This is of great importance because the benefit is felt by almost all components of the entire ecosystem. If humans exercise waste minimization with energy and dedication, it may prevent the earth from suffocating in all the wastes that humans generate.

Improved Employee Safety

There will be fewer hazardous materials and wastes for employees to handle in an organization if waste minimization is practiced. Reduced handling equates to less exposure and improved employee safety.

Regulation Compliance

Waste minimization practices allow compliance with several federal regulations. To comply with RCRA, the generator must certify that they are implementing waste minimization. There is some waste minimization reporting associated with SARA Title III. Biennial reports require a description of the efforts undertaken during the year to reduce the volume and toxicity of waste generated. Waste minimization will help organizations achieve compliance with federal and state land disposal bans. Finally, TSDF permits require that the generator of the hazardous waste have a program in place to reduce the volume or quantity and toxicity.

Two more recent important federal bills include the Wolpe Schneider Bill, or Waste Reduction Act of 1989 (HR 145), and the Lautenberg Bill, or Pollution Prevention Act of 1990. To comply with this latter act, all facilities are required to submit, with their Form Rs, an annual toxic chemical source reduction and recycling report. The report must include: the amount of wastes entering the waste stream and percentage change from the previous year; the amount of wastes into the waste stream expected to be reported for the next two calendar years; the amount recycled, percentage change in amount recycled from the previous year and process used; source reduction practices used; ratio of the reporting year's production to the previous year's production; techniques used to identify source reduction opportunities; and the amount of toxic chemical released as a result of one-time events.

California Senate Bill 14 is resulting in dramatic changes in companies and agencies concerning waste minimization, especially source reduction. As of September 1, 1991, all generators of 12,000 kg/yr or more of hazardous waste or 12 kg/yr of extremely hazardous waste have to submit several reports. In order to prepare the reports, new organizational programs were needed in most companies. These reports include Source Reduction Evaluation Review And Plan, Source Reduction Evaluation Review And Plan Summary, Hazardous Waste Management Performance Report, and Hazardous Waste Management Performance Report Summary. California SB 1726 in 1992 broadened the hazardous waste minimization requirements and included more small-quantity generators.

For each hazardous waste stream identified, the plan must include an estimate of the quantity generated and an evaluation of the source reduction methods available to the generator. In addition, the plan must include reasons for the reduction methods selected and a full explanation for the rejection of methods. Last, an evaluation of the effects of the chosen method and a timetable for adoption must be included.

Operating Cost Reduction

After some possible up-front expenditures, waste minimization usually results in operational cost savings. The greatest savings will be in less HMW storage, transportation and disposal costs. This is significant since disposal costs have been increasing by as much as 50%/year. There is also cost savings due to less raw materials used and less manpower needed (smaller legal, regulatory and environmental staffs, for example). Finally, there may be tax advantages and lower insurance costs.

Improved Public Relations

Improved public relations is a big benefit of waste minimization that's hard to quantify. Public relations has been known to make or break an organization, especially in the hazardous waste arena. 3M and Dow Chemical are two examples of companies that have received considerable positive recognition for their programs.

Most waste minimization efforts can lead toward Green Certifications, which are becoming popular in many parts of the world. Europe leads with programs such as the Green Point, Eco Mark, Blue Angel and White Swan. The U.S. has the Green Cross and Green Label programs. All of these programs are voluntary and oriented toward marketing, public relations and waste minimization.

Reduced Liabilities

Waste minimization can help to minimize future cleanups, court action and fines for noncompliance. As with public relations, this benefit is also hard to quantify and can be in the multimillions of dollars [Higgins 1989]. If there are fewer HMWs present, there is less chance of problems leading to fines and law suits.

Generic Waste Minimization Program Key Elements

The following list presents some of the major steps in setting up a waste minimization program. Depending on the number and complexity of waste streams, fewer or additional steps may be required. Large waste minimization projects should be carried out by a team. The team needs to be identified early, at least by step 5. If a team is utilized, it is essential to assign responsibilities. Steps 2, 4, 9, 10 and 13 are especially important and are designated by the EPA as "Effective Program Elements."

1. *Set and Prioritize Objectives*—Take care to clearly state a manageable number of realistic objectives. Most establish too many goals and then "shotgun" the team's energy in too many directions. This can lead to none of the objectives being accomplished.

2. *Perform an Initial Waste Stream Reduction Audit*—The audit should include the types, amounts and level of wastes generated and the sources of the wastes. In other words, don't set everything up from a conference room or a desk. Physically get out to where the wastes are being generated and learn about their evolution.

3. *Identify and Prioritize Waste Streams to Minimize*—The prioritization should consider what is feasible and what is not. In identify-

ing the waste streams to pursue, involve all employees, not just environmental compliance staff. For example, include the workers and managers involved in production and maintenance.

4. *Obtain Top Management Support*—The president of the organization must support waste minimization verbally, in writing and financially. Without this support the program is doomed. It is also a good idea to get a show of support from the senior vice presidents as well.

5. *Perform Periodic On-Site Assessments*—By routinely going out in the field and reviewing operations, new waste streams will become obvious candidates for minimization.

6. *Involve, Motivate and Educate All Employees*—In addition to including as many people in the identification phase as possible, the entire organization should be involved in most other phases of waste minimization. For example, reward programs will encourage employees to think and support new ideas. Many companies have found that dollar awards, plaques, and honorable mentions in company newsletters are all good ways to increase employee involvement [Higgins 1989].

7. *Design and Evaluation of Action Plans*—An action plan needs to be carefully prepared. This may even include some design. A technical and economic evaluation of the action plans would then occur and result in the selection of a manageable number of realistic action plans to be implemented.

8. *Testing of Selected Action Plans*—A test run of the action plan(s) is advisable. It's easier to get the bugs out on a small scale.

9. *Obtain Funding*—Many good projects have come to a screeching halt because of a lack of funds. The funds should be firmly committed before much energy is expended.

10. *Revise Accounting Methods and Distribute*—Certain waste minimization practices will require changes in the organization's account-

ing system. If this is identified as being needed, it should be started early, since most accounting systems take considerable time to change. Accounting systems must be in place to allow for the different ways in which the hazardous materials and wastes will be handled and tracked from a financial standpoint.

11. *Revise Operational Procedures and Distribute*—As with accounting, sometimes methods and procedures will need to be written or rewritten to include the new waste minimization components. They must be written clearly and in terms that all affected individuals can understand. Many refer to these instructions as "best management procedures or practices."

12. *Implement the Waste Minimization Actions*—This is what it's all been building up to and what all the planning is about. Now it's time to physically start the new system or process.

13. *Evaluation Program and Follow-Up*—It is important to ensure that the waste minimization techniques are being utilized since old habits are sometimes hard to break. Follow-up visits and memos will help ensure this.

In July 1988, the EPA published its Waste Minimization Opportunity Assessment Manual. In this document, the EPA basically identified four steps: 1. Planning and organization (see steps 1–4 above); 2. Assessment phase (see step 5); 3. Feasibility analysis phase (see steps 7 and 8); 4. Implementation (see step 12).

Waste Stream Ranking Considerations

Some operations can involve hundreds of hazardous waste streams. It may be hard to decide which one(s) to concentrate on first. The following considerations may help in this determination and are arranged roughly in order of importance, depending on your perspective: environmental and employee safety impact, feasibility of implementation, regulatory compliance (land bans), effectiveness of the waste minimization action, disposal costs, production and quality risk, monthly amount produced, operating and capital costs, and investment potential [Higgins 1989]. Volume of waste generated and exposure time are indirectly covered in the above categories.

The environment and employee safety impact component mentioned above can itself be subdivided into several categories. One important component is a toxicity rating, which commonly uses LD50. Closely related is the waste's carcinogenicity, mutagenicity or teratogenicity. The waste's flash point and reactivity are also important considerations.

Waste Minimization Options and Priorities

The waste minimization options could in general be applied at the manufacturer, distributor, storeroom, user and contractor's facilities. A few of the options may, however, be a little more practical at one phase of the HMW flow than another. Most of these options could also be applied to industry, households and the other types of generators. The following options are presented in order of priority and grouped into either hazardous material minimization or into hazardous waste reduction categories.

Hazardous Material Minimization (Source Reduction)

The heart of HM minimization is source reduction. If possible, the HM should not be allowed to enter the operation in the first place; otherwise, minimize the amount needed. HM minimization can be achieved through the following overlapping options.

Substitution

If at all possible, replace products or raw materials that contain HM for those that don't. If that doesn't work, replace with ones that contain less HM in terms of volume or toxicity. Substitute processes and equipment that use HM for those that do not [Higgins 1989].

New process or equipment substitution is the most desirable form of waste minimization. You simply design a new system, process or piece of equipment that does not contain or use hazardous materials. This is usually harder than it sounds, but definitely achievable.

Existing process or equipment substitution is the next best thing to do. Some of the HM may already be in the system. By making substitutions, changes and upgrades you can possibly phase out the HM or at least minimize it. There is usually some resistance from operational personnel when this approach is taken since they are already dependent on the HM. Sensitivity and planning are therefore necessary.

Figure 14–2 is an example of the process taken by the author to introduce a lesser toxic industrial adhesive. The first half of the figure illustrates the steps used to select the hazardous material to minimize. In this case, it turned out to be an industrial adhesive. The second half of the figure illustrates the screening process used to select the best substitute for the presently used industrial adhesive.

Inventory Control

By buying only what is needed at the time and rotating, the inventory of HM can be kept at a minimum. Also, if the number of different brand names can be reduced, it will help keep the inventory at a lower level.

Purification of Raw Materials

Sometimes, if one or more of the raw materials can be purified before being fed into the process, it will generate less waste. The manufacturer should be encouraged to clean up its product to higher quality standards. The purchaser may still need to purify the raw material even further. In addition to reducing hazardous waste, this may even improve the quality of your own product or service.

Development of New Operating Procedures

It is possible to write procedures that will help minimize HMW. For example, an operating practice to screen MSDSs for all new products purchased or about to be purchased would probably help minimize the amount of HMW the organization must contend with. These procedures are also known as management practices, standard operating procedures and other terms.

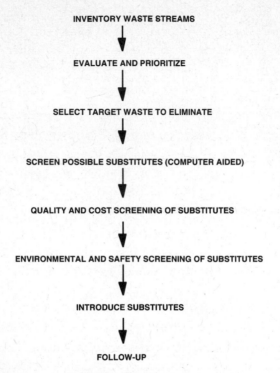

Figure 14–2 Waste Minimization Process Example.

Production Scheduling

Tightening up the production schedule, so that there are fewer start-ups and shutdowns will minimize waste. This is possible because more waste is usually generated when the machines (and operators) are cold. Alternating operating schedule and rearranging work shifts might minimize cumulative product waste, emissions and risks.

Better Housekeeping

Housekeeping was also discussed in Chapter 9. Keeping areas clean, in order and with a minimum of waste present is an excellent way to facilitate all of the other waste minimization options. Labeling and segregating HMW are also good housekeeping practices.

Hazardous Material Reuse, Exchange or Sale

Some HM can be reused within the same organization, even without treatment. For example, paint thinner can be reused many times, merely by allowing a few hours of settling time. Also, many high-grade solvents used in an ultraclean manner may be reused in areas with less stringent quality standards.

If some effort is expended, it may also be possible to find someone outside of the organization who may want to have the HW. Waste exchanges help facilitate this by listing unneeded HMW from various organizations. Waste exchanges were developed in Europe and are starting to become established in the United States. If the HMW is exchanged, donated or sold, it might not even be considered a waste since it has a functional use in its present form. It is recommended that when this occurs, a bill of sale is prepared, even if it is for $1. When something is sold, considerable responsibility (but not all) passes to the new owner.

Improved Efficiency

If the entire flow of HM is made as efficient as possible, HM can be minimized. Efficient manufacturing, distribution, storage and use will aid in the effort. Manufacturing and materials handling are two areas where efficiency improvements will have a big impact.

Better Equipment Maintenance and Monitoring

Maintenance and monitoring are closely related to improved efficiency. If equipment utilizes or produces HMW, it should be monitored and maintained. This will improve the quality and minimize the amount needed. Improved maintenance will help in reduction of spills, leaks and vaporization.

Improved Mass Balance Tracking and Product Conservation

Also related to the above is mass balance tracking. Complicated processes may have interrelated hazardous material streams. It is essential to accurately account for all of the material at all times. Among other things, this will help conserve the raw materials.

Altering the Process

By adjusting the process it is possible to eliminate or reduce hazardous material end products and byproducts. This is closely related to the section already presented, "Development of new operating procedures" page 239. For example, using physical or mechanical cleaning (such as hydroblasting, rodding, brushing, wiping and scraping) in conjunction with solvents will reduce HW. These are modifications that result in less chemical use or safer chemicals used.

Hazardous Waste Reduction

If it is not possible to further minimize or reduce the hazardous materials at the source, HW reduction should be implemented. At this point, we are concentrating on reducing the total volume or toxicity of waste that has already been generated.

Recycling, Reclaiming and Reuse

Certain processes allow for reuse of part of the waste stream after some treatment. For example, organic solvents, antifreeze, silver, sulfuric acid and oil are commonly recycled. If possible, on-site recycling is preferred over off-site.

Assuming you can't minimize the HM at the source, reuse and recycling is the next best option. This may require permits and still generate some waste, but at least part of the waste stream will be reutilized, thus minimizing the amount of new HM needed and the amount requiring disposal.

Recycling is one of the easiest options to recognize. When most individuals refer to waste minimization, they are normally using recycling examples. Waste oil, antifreeze and batteries are recycled by many people. Most industries are now recycling solvents. Specific industries are even recycling unique waste materials, such as waste sulfuric acid.

Most recycling processes have some economic limitations on the type and quantities that can be recycled. The market for some recycled materials is acceptable, while other markets are flooded. This fact determines how much energy and dollars can be spent to recycle different wastes. The cost of straight disposal is also a consideration.

Recycling is done either on-site or off-site. If the volumes allow it to be done economically on-site, transportation costs can be avoided. Sometimes a TSDF permit is required whenever any new hazardous waste is created from the recycling process. In some states there are some exceptions, variances and alternate permit options available that avoid the requirement for a full-blown TSDF permit. Also, if the hazardous material is pulled out of the process and recycled before it becomes a waste, and the recycle process does not generate a hazardous waste, a TSDF permit may not be required. In this situation, it would be considered a closed loop.

Recycling in many cases basically involves the removal of impurities and the addition of additives to functionally restore the material. The removal step involves processes that separate dirt, metals, bacteria and other contaminants from the product. Examples of removal processes include settling, filtration, straining, cyclone separation, centrifugation, magnetic separation, pasteurization/distillation and ultrafiltration. Once the contaminant is removed, the material is used as is or functionally restored by adding conditioners, emulsifiers, concentrates, surfactants, bactericides, antioxidants and other chemicals.

Many factors must be considered when deciding whether to recycle and, if so, with what system. Probably at the top of the list are the economic factors such as present disposal cost, and type and quantity of waste. Equipment effectiveness, operational simplicity, maintenance, floor space requirements and, of course, degree of environmental impact reduction are also important [Higgins 1989].

Improved Waste Tracking

As with HM tracking, it is essential to have an accurate mass balance for HW. The manifest system is an integral part. However, prior to manifest preparation, the generator can help minimize waste by better waste tracking. This would involve knowing where all effluents, emissions, bulk and containerized waste are sent.

Waste as a Fuel or Construction Material

In certain cases and with agency approval, some HW can be used as a fuel (such as in cogeneration for energy recovery), as a construction material additive (such as in concrete) and for other uses. Some states consider the burning of used oil as recycling. In terms of construction additives, 40 CFR 248 requires that federal agencies procure building insulation materials that contain recovered materials, and 40 CFR 249 requires that concrete products contain a minimum amount of fly ash. Care must be taken, however, to ensure that the waste is adequately "fixed" and will not leach toxics into the environment if it is used as a construction material. If it is used for energy recovery, it is especially important to make sure the contaminants are completely destroyed during the process.

Treatment

Treatment will minimize the volume or toxicity of the waste requiring disposal. Unfortunately, treatment often requires numerous permits, which is usually expensive. Treatment should only be considered if the other waste minimization options are completely unacceptable.

Waste Minimization Assistance

The EPA and many state agencies are offering technical assistance in waste minimization. For example, information sources, grants and loans are available for waste reduction. In addition, the California Waste Exchange promotes waste reduction and exchange.

Appendix F
ISO Templates

Suggestions and Tips
On How to Best Use the Disk

ISO 14001 environmental management procedure templates are found in the floppy (and in this appendix). The floppy is formatted in Microsoft Word for Windows 6.0 and Word Perfect 5.1.

The floppy includes three files of general information followed by 19 specific procedure templates. The general information files include 1-tips.doc (this page), 2-index.doc, and 3-format.doc. The 19 specific template files containing environmental management procedures are arranged alphabetically on the floppy. By using these files the reader will not have to start from scratch. The 19 template files contain the major procedures that will be needed to obtain an ISO 14001 certification. They are written generally so that they will apply to any organization worldwide. The reader should insert their own site-specific information into these procedures.

For individuals unfamiliar with computers and software the following suggestions are made to get started with a minimum of pain and suffering:

1. **Word Processing Software** – Load your word processing software. It must be one of the following: Microsoft Word for Windows 6.0 or later or Word Perfect 5.1 or later.

2. **Floppy Included with This Book** – Load the floppy included with this book.

3. **Floppy Software Files** – Call up the floppy software file which corresponds to the software in your computer (i.e., Microsoft Word or Word Perfect).

4. **General Information Floppy Files** – Select and read the files in the following order: 1-tips.doc (this page); 2-index.doc; and 3-format.doc.

5. **Specific Procedure Floppy Files** – Select the template for the procedure(s) on which you want to work. There are 19 files arranged alphabetically according to title. Discussion concerning the files can be found in the book and is presented in logical sequence. Each floppy file ranges from four to six pages.

6. **Backup Floppy** – Make a backup copy of the floppy before you start using these templates.

7. **Use of the Templates** – Add your own site-specific information and delete what doesn't apply.

8. **Backup New Additions** – Backup what you add.

Cross Reference Summary of ISO Templates Presented in this Book

Word Processor Template Index Name	Appendix F – Title of Procedure	Applicable Chapter

*Initial Word Processing Guidance Files**

1-tips.doc or .wp	Suggestions on How to Use the floppy	
2-index.doc or .wp	Cross-Reference/Index (this page)	
3-format.doc or .wp	Blank Template	

Boiler Plate Environmental Procedure Templates

approval.doc or .wp	Process, Equipment and Chemical Approvals	12
auditing.doc or .wp	Audit, Review and Verification Procedure	14
awarenes.doc or .wp	Internal Communication & Employee Awareness	7
chemcont.doc or .wp	Handling of Empty Chemical Containers	12
chemtrck.doc or .wp	Tracking of Chemicals	12
closure.doc or .wp	Site Closure	21
communic.doc or .wp	Procedure for External Communication	17
data coll.doc or .wp	Data Collection and Handling	15
disaster.doc or .wp	Disaster Recovery Plan	13
discharg.doc or .wp	Minimizing Discharges to Air and Water Bodies	19
emergenc.doc or .wp	Emergency Procedures	13
finances.doc or .wp	Identification and Tracking of Finances	7
impacts.doc or .wp	Dealing with Environmental Impacts	8
objectiv.doc or .wp	Identification of Objectives and Targets	10

(Continued on next page)

---

*Microsoft Word for Windows files end with a .doc extension, and Word Perfect files end with a .wp extension.

TITLE:	PROCEDURE NO:	Page _____ of _____
Company Name:	**Date:**	*Rev.*

Company Name: _____

Site: _____

This is a "controlled" document.
Routine distribution is restricted to the approved
distribution in _____. All other persons
in possession of this document have uncontrolled copies
and should call document control for revision level status.

Approved by _____

Approved Date _____

Confidential

TITLE:	PROCEDURE NO:	Page _____ of _____
Company Name:	Date:	Rev.

Procedure for _____

1. Purpose

1.1.

1.2.

2. Scope

2.1.

2.2.

3. Responsibilities

3.1.

3.2.

4. Procedure

4.1.

4.2.

5. Related Documentation

TITLE:	PROCEDURE NO:	Page _____ of _____
Process Equipment and Chemical Approval Procedure		
Company Name:	**Date:**	*Rev.*

Company Name : _____

Site: _____

Process, Equipment and Chemical Approval Procedure

This is a "controlled" document.
Routine distribution is restricted to the approved
distribution in _____. All other persons
in possession of this document have uncontrolled copies
and should call document control for revision level status.

Approved by _____

Approved Date _____

Confidential

TITLE:	PROCEDURE NO:	Page ____ of ____
Process Equipment and Chemical Approval Procedure		
Company Name:	**Date:**	*Rev.*

Process, Equipment and Chemical Approval Procedure

1. Purpose

1.1. To insure that the environment is considered during the design process in order to minimize environmental impacts.

2. Scope

2.1. This procedure applies to design processes that specify chemical or natural resource usage, generation of hazardous and nonhazardous waste, discharges and other impacts to the environment.

3. Responsibilities

3.1. The design department is responsible for notifying the environmental department of new designs which are within the scope specified above.

3.2. The environmental department is responsible for timely evaluation of the designs and cost-effective recommendations.

4. Procedure

4.1. Advance Memo

Memos will be sent in advance to all departments that might implement a new chemical or process or install a new piece of equipment. The department will be asked to send a copy of any drawing or design document for review as soon as a draft is available. See Figure 12–2 for an example. This particular memo should be sent several times, as a reminder.

TITLE:	PROCEDURE NO:	Page _____ of _____
Process Equipment and Chemical Approval Procedure		
Company Name:	Date:	*Rev.*

4.2.   Narrowing of Request

The environmental manager should not be surprised if all drawings are not sent. Even 50 percent is optimistic in some organizations. If resistance is encountered, and it may be, qualify the above request by asking for designs that meet the following considerations:

- Uses a chemical in the process
- Discharges to the air, water or land
- Generates a hazardous waste
- Uses a chemical in the maintenance of the process or equipment
- Generates a solid waste such as paper, aluminum, glass or plastic
- Utilizes considerable energy, water or other natural resource

4.3. Review of Materials

When the drawing or design arrives it will be reviewed immediately to ensure compliance with environmental, health and safety requirements. If not enough information has been sent, a request for additional information (see Figure 12–3) will be sent.

4.4.   Field Review

If anything of concern is found in the paperwork, a personal visit to the design engineer and location should be made to analyze the situation in greater detail. The field visit should be made if at all possible independent of what the paperwork review uncovered.

4.5.   Recommendations

Suggestions will be made in writing concerning ways to minimize the proposed use of the chemicals, natural resources, emissions or the impacts. Waste minimization concepts, engineering and administrative controls and personnel protective equipment are all possible approaches to consider.

TITLE:	PROCEDURE NO:	Page ____ of ____
Process Equipment and Chemical Approval Procedure		
Company Name:	Date:	*Rev.*

4.6. Sign-off

Sign-off or corrections of the proposed chemical, process or equipment will be made by the environmental, health and safety department (see Figure 12–4). Notify the requester, the purchasing department and the receiving department of the sign-off and/or conditions.

4.7. Ordering

When it is time to order the chemical or equipment, it is necessary to work closely with the purchasing department. This will allow one additional layer of screening and control to ensure that only authorized vendors are used. In most cases the environmental manager starts to lose control and contact at this point since the chemical or equipment is usually ordered by someone else in the organization.

4.8. Follow-up

Because of the problem just mentioned it is essential for the environmental manager to followup. Follow-up will be scheduled to ensure the recommendations were followed and that the overall design for the process or equipment has not changed radically.

5. Related Documentation

5.1. Approval Signatures on File

5.2. Conceptual, Preliminary and Detailed Design Drawings

TITLE:	PROCEDURE NO:	Page _____ of _____
Audit, Review and Verification Procedure		
Company Name:	**Date:**	***Rev.***

Company Name: _____

Site: _____

Audit, Review and Verification Procedure

This is a "controlled" document.
Routine distribution is restricted to the approved
distribution in _____. All other persons
in possession of this document have uncontrolled copies
and should call document control for revision level status.

Approved by _____

Approved Date _____

Confidential

TITLE:	PROCEDURE NO:	Page ____ of ____
Audit, Review and Verification Procedure		
Company Name:	Date:	*Rev.*

Audit, Review and Verification Procedure

1. Purpose

1.1. To continuously improve the environmental management and control systems in order to minimize impacts to the environment.

2. Scope

2.1. This procedure applies most directly to the company's own operations.

2.2. This procedure also applies to the organization's suppliers or vendors.

3. Responsibilities

3.1. It is the responsibility of the environmental department to conduct the audits or to arrange for a third-party audit.

3.2. The procurement department is to assist in terms of arranging audits of vendor facilities and in contracting with third-party auditors.

3.3. The operation departments have the responsibility to promptly correct audit findings or provide documentation why an audit correction is not necessary.

4. Procedure

4.1. Initial Contact

The concept of the audit will be introduced to the site along with any available information. Some people react negatively to the word audit, so it might be best to refer to it as an assessment or something less onerous

TITLE:	PROCEDURE NO:	Page _____ of _____
Audit, Review and Verification Procedure		
Company Name:	Date:	*Rev.*

4.2. Schedule and Purpose

Schedule the audit date with the site and alert them to the purpose of the visit. Try to accommodate their production needs. Point out that one purpose of the assessment is to prevent agency mandated shutdowns due to regulatory noncompliance.

4.3. Pre-Audit Conference

Upon arriving at the site there will be a pre-audit conference to go over the plans for the day. All questions that the operation personnel have should be adequately addressed. An attempt to solicit and cover special requests made by site personnel during the normal audit should be made.

4.4. Review the audit checklist

A list of what to look for, such as in Appendices C and D, should be reviewed. It is impossible for the auditor to remember all issues that should be checked. An actual audit list would be two or three times as long since it should also include local regulatory issues and organization specific policies.

4.5. Site Inspection

The next step will be to walk the site and record observations in a factual way. Also note items in a way that minimizes liabilities, if possible. For example, a record which creates liability problems might be: A hazardous waste sign is missing at the drum storage area. Instead record the same observation in the following way which has fewer liability problems: Hazardous waste signs are required at hazardous waste storage areas. Verify that one is in place.

4.6. Post-audit Meeting

A post-audit meeting will be held to discuss initial findings. It is usually not a good idea for the auditors to insist on agreement concerning corrective actions on the spot. This applies whether the auditor is internal or external to the organization. To expect immediate agreement and corrective action commitment is not reasonable and might even damage the relationship with the site operating personnel.

TITLE:	PROCEDURE NO:	Page _____ of _____
Audit, Review and Verification Procedure		
Company Name:	Date:	*Rev.*

4.7. Analyze Findings

In order to make meaningful statements, the auditor will analyze the audit findings and research applicable regulations. Corrective actions that are realistic and cost-effective should be suggested. Make sure that the corrective actions are really required by regulations or company policy.

4.8. Enter Data

Entry of the audit findings into the database will be done along with suggested corrections and if possible a reference to a specific code, standard or regulation. Give a draft to the site individuals. There are numerous data base management software systems available. Some, however, are limited in capacity, sorting and analysis ability, which should be kept in mind when purchasing hardware and software.

4.9. Transmit Audit

The actual audit will be sent out to the lead site environmental, health and safety individual (and "cc" senior management on memo). Only individuals with a need to know should receive the audit or liability problems will be created. Also if the actual audit goes to upper management, there may be resentment by the individuals who have to implement the actions. In addition it is not to the best interest of upper management to concern themselves with details, such as a missing label, unless it is not corrected in a reasonable time frame.

4.10. Follow-up

A corrective action reminder will be transmitted if the site has not responded with a correction. It may be necessary to do this several times until all open audit items are closed. Follow-up is when most audit systems fall apart. Usually more than enough problems and corrective actions are identified, with only a portion of them being completed.

TITLE:	PROCEDURE NO:	Page ____ of ____

Audit, Review and Verification Procedure

Company Name:	Date:	*Rev.*

4.11. Continual Improvement of Audit System

It is important to adjust the audit checklist to ensure that it is meaningful and comprehensive. The audit team composition and other components of the entire audit system may need to be adjusted.

4.12. Corrective Action Procedure for Noncompliant Items

When action items are noted during an audit they will be corrected quickly. An organization should not create "smoking guns" which are noncorrected audit items noted in the file. In order to make sure correction is done quickly and efficiently, the following procedure will be followed:

4.12.1. Identification

The first obvious step is to identify the action that needs correction. It is better from a liability standpoint to record the observation as corrective actions in place of a list of noncompliant items.

4.12.2. Method of Correction

The suggested corrective action may include an action plan. Most individuals being audited would appreciate some feasible suggestions, not just a list of problems.

4.12.3. Transmittal

Send the corrective action required to the party who is responsible for implementation of the correction. There may be many people who actually correct a finding, therefore one person with overall responsibility for correction should receive the notice.

4.12.4. Followup

Followup to ensure that the correction has been completed. Depending on the item, this may require another site visit, a call and/or several memo reminders. The frequency of the follow-ups is determined by the seriousness of the deficiency. A form should be sent to the site at least once per quarter until the audit item has been corrected.

TITLE:	PROCEDURE NO:	Page _____ of _____
Audit, Review and Verification Procedure		
Company Name:	Date:	*Rev.*

4.12.5. Continuous Improvement

It may be necessary to change an overall system or procedure that may affect several sites. The audit procedure itself may need to be changed. Continuous improvement is important here as well as in most other aspects of environmental management.

4.12.6. Audit Documentation

Each site will have the ability to document their progress and results in terms of the audits. Much data that is generated is the result of an audit recommendation. For example, monitoring and other test data are often specified in an audit. In order to complete the loop or close an open audit item, the audit documentation must be in the file. Since the audit documentation is so important, quality control is necessary.

4.13. Audit Reporting Procedure and Management Review

The findings and recommended corrective actions from audits will be relayed in writing to the site environmental manager. Usually it is politically advantageous to give the site environmental manager the only copy to provide them with an opportunity to correct many of the items before senior management becomes involved. If the site environmental manager facilitates correction of the significant findings promptly and completely, it may not even be necessary to give senior management the complete audit. In this case at least the cover memo should be sent to senior management to let them know that an audit has been done and a phone number if they want a copy. The memo should also state that all significant discrepancies uncovered by the audit have been corrected by on-site personnel, if this is the case.

Management should use the audit findings to their full extent. This is valuable information that can help the organization improve. It is also information that can lead to liability problems if it is not acted upon but is documented.

Targets, objectives and policies may be adjusted based on the audits. When and if this is done, the site management must be involved.

TITLE:	PROCEDURE NO:	Page _____ of _____
Audit, Review and Verification Procedure		
Company Name:	**Date:**	*Rev.*

4.14. Continual Audit Improvement

After the programs, process and plans discussed are implemented they will be continually improved. The improvement would be based on data obtained, audit results and other types of feedback available. The audit program will be reegineered or continually adjusted based on output and input from other systems.

5. Related Documentation

5.1. Audit Report

5.2. Audit Criteria

5.3. Statements Why an Audit Correction Is Not Required

5.4. Date of Closure of Audit Items

TITLE:	PROCEDURE NO:	Page _____ of _____
Internal Environmental Communication and Employee Awareness		
Company Name:	Date:	*Rev.*

Company Name:_____

Site: _____

Internal Environmental Communication and Employee Awareness

This is a "controlled" document.
Routine distribution is restricted to the approved
distribution in _____. All other persons
in possession of this document have uncontrolled copies
and should call document control for revision level status.

Approved by _____

Approved Date _____

Confidential

TITLE:	PROCEDURE NO:	Page ____ of ____

Internal Environmental Communication and Employee Awareness

Company Name:	Date:	*Rev.*

Procedure for Internal Environmental Communication and Employee Awareness

1. Purpose

1.1. To build understanding, cooperation, buy-in and involvement concerning environmental protection by as many employees as possible.

2. Scope

2.1. This applies to all employees of the organization, no matter what level.

2.2. This applies to on-site contractors of the organization.

3. Responsibilities

3.1. Communication of environmental issues is the responsibility of the environmental department.

3.2. It is the responsibility of every employee to communicate environmental issues, problems or ideas to the environmental department.

4. Procedure

4.1. Communication through Normal Management Channels

The traditional flow of information from one management level to the next is appropriate and will be used for certain environmental information. For example, managers will relay to their employees that they should come forward with waste minimization ideas. The managers will instruct their employees to follow published procedures for empty chemical container handling, for example.

TITLE:	PROCEDURE NO:	Page _____ of _____
Internal Environmental Communication and Employee Awareness		
Company Name:	Date:	*Rev.*

4.2. Communication though Internal Newsletters

Certain environmental communications will be made through the internal company newsletters. For example, the residents signing a new environmental policy on chemicals, such as Class I ozone depleting substances, would make a good story. All employees should be aware of events of this magnitude and feel motivated and proud. Stories of this importance should also be considered for public release. This will build confidence in the company and raise employee morale.

4.3. Communication through Videos

Videos will be used to relay environmental information to large groups of employees. If the video is endorsed in writing or verbally by a senior officer the impact of the video can be maximized. Due to cost and attention span, the length of videos will not exceed 20 minutes.

4.4. Communication through Desk Drops, E-mail, Posters and Special Memos

Announcements concerning waste minimization programs, upcoming agency audits and other general interest information will be disseminated though desk drops, e-mail, posters and memos to large populations of employees. For this to be an effective method to communicate environmental information, it will be done frequently. It should be kept in mind that most employees are busy with producing the product or service and may not pay much attention to some of this type of communication.

4.5. Communication through Suggestion Boxes and Employee Hot Lines

Most of the communication methods mentioned above are from the environmental management individuals or department to the employees. In the case of suggestion boxes and employee hot lines, the communication is coming from the employees. Employees are a valuable source of information about actual impacts that are occurring in the field. They will also be the ones to come up with most of the realistic waste minimization ideas. Therefore this communication

TITLE:	PROCEDURE NO:	Page _____ of _____
Internal Environmental Communication and Employee Awareness		
Company Name:	**Date:**	*Rev.*

channel will be made readily available for use by everyone. Rewards will be considered for suggestions. Employees must also be told that they need to report suspected violations to designated individuals in the organization.

4.6. Special Communications via Attorney

Certain environmental information should be relayed only to those with a need to know by an attorney. When communication is done in this way the information is protected by the Attorney-Client Privilege doctrine. This means that if the whole situation ends up in court, the information would not have to be discovered or brought forward, which is important if it is supposition, guess work or theories.

4.7. Recognition Communication

Both positive and negative recognition must be given to employees. This will help minimize the chance of fines and prison. The positive recognition could be in the form of awards, certificates and prizes to reward employees for waste minimization ideas, for example. Negative recognition must be given if employees knowingly violate regulations or company policies which negatively impact the environment. The negative recognition is usually in the form of verbal and written warnings, suspension or termination.

5. Related Documentation act

5.1. File Containing Internal Communications Sent to Employees

5.2. File Containing Employee Feedback and Action Taken to Address Feedback

TITLE:	PROCEDURE NO:	Page _____ of _____
Handling Empty Chemical Containers		
Company Name:	Date:	*Rev.*

Company Name:_____

Site: _____

Handling Empty Chemical Containers

This is a "controlled" document.
Routine distribution is restricted to the approved
distribution in _____. All other persons
in possession of this document have uncontrolled copies
and should call document control for revision level status.

Approved by _____

Approved Date _____

Confidential

TITLE:	PROCEDURE NO:	Page ____ of ____
Handling Empty Chemical Containers		
Company Name:	Date:	*Rev.*

Procedure for Handling Empty Chemical Containers

1. Purpose

1.1. To ensure that chemical containers are properly handled so that they do not impact the environment or injure employees.

1.2. To encourage the reuse and recycling of empty chemical containers.

2. Scope

2.1. This procedure covers any empty container or one with residue that housed a chemical that might have an impact on the environment or human health.

2.2. This procedure covers all types of hazardous chemicals or fuels whether they are in the liquid, solid or gaseous/vapor state.

3. Responsibilities

3.1. It is the responsibility of the chemical user to ensure that the chemical container has been properly cleaned.

3.2. The environmental department will audit the container cleaning operation to ensure that the procedure is being followed.

3.3. The procurement department will arrange contracts with chemical suppliers to reuse as many containers as possible.

4. Procedure

Since most empty chemical containers may still have some residue, they must be handled properly to minimize injury and impact to the environment. In many cases, the residues in the containers result in a hazardous waste classifica-

TITLE:	PROCEDURE NO:	Page _____ of _____
Handling Empty Chemical Containers		
Company Name:	Date:	*Rev.*

tion. Organizations should have procedures for the handling of empty chemical containers, whether they are common chemicals, such as photocopy toners, or the highly toxic chemicals.

4.1. Training

Train employees concerning this procedure and provide personal protective equipment (PPE) for when it is carried out. If the employees don't understand or appreciate what they are doing, there will be containers going out in the trash with residual chemicals. This may result in heavy fines and possible environmental damage.

4.2. Complete Use

It is very important to remove as much of the chemical as possible. This may require inversion for a period of time or scraping or chipping. Drain racks are commonly used for this phase of the process.

4.3. Handling of Residual

The residue will be used in the plant operation if possible. If this cannot occur then the residue will probably have to be treated as hazardous waste and should not be placed in the common sanitary landfill. If liners are present, they are also usually treated as hazardous waste.

4.4. Container Cleaning/Washing

Depending on the residue and whether the company has washing and treatment equipment, it may be possible to wash out the container and cap. This washing is usually done three times (triple rinse). The wash solution may have to be treated as hazardous waste.

TITLE:	PROCEDURE NO:	Page _____ of _____
Handling Empty Chemical Containers		
Company Name:	**Date:**	*Rev.*

4.5. Testing

The container will be tested after the washing. For example, if an acid or base was in the container before, then pH paper could be used. pH meters are another option to ensure adequate washing has occurred.

4.6. Inversion

After washing the container, it should be inverted. A drain rack will allow all residual fluid to exit the container.

4.7. Return of Container

If possible, the container should then be returned to the chemical supplier, broker or recycler. This is preferable to land disposal in terms of the environment and liabilities.

5. Related Documentation

5.1. Receipts From Vendors Showing Pickup of Empty Container Destined for Reuse or Recycling

TITLE:	PROCEDURE NO:	Page _____ of _____
Handling Empty Chemical Containers		
Company Name:	**Date:**	*Rev.*

Company Name:_____

Site: _____

Tracking of Chemicals

This is a "controlled" document.
Routine distribution is restricted to the approved
distribution in _____. All other persons
in possession of this document have uncontrolled copies
and should call document control for revision level status.

Approved by _____

Approved Date _____

Confidential

TITLE:	PROCEDURE NO:	Page ____ of ____

Handling Empty Chemical Containers

Company Name:	Date:	*Rev.*

Procedure for Tracking of Chemicals

Tracking of chemicals is an essential component of a comprehensive environmental management system. Being aware of chemicals which are planned, purchased, stored and used and which become hazardous wastes is the essence of the system. Key components of the tracking system include an inventory along with important information about the chemical and a mass balance.

If some of the same chemicals are used at different locations for the organization, it might be more cost-effective to have a central or corporatewide software system. This system would, however, have to have a process where site-specific information could be easily entered.

1. Purpose

1.1. To make sure all chemicals are identified and accounted for in all phases of the operation.

1.2. To reduce the volume and/or toxicity of chemicals on-site.

1.3. To detect early any significant leak or spill of chemicals.

2. Scope

2.1. This procedure covers chemicals used in R&D, preproduction, manufacturing and all other aspects of the operation.

3. Responsibilities

3.1. The purchaser of the chemical is responsible for notifying the environmental department when a new chemical is introduced into the operation. The purchaser is also responsible for tracking of chemical quantities and reporting this to the environmental department.

TITLE:	PROCEDURE NO:	Page _____ of _____
Handling Empty Chemical Containers		
Company Name:	Date:	*Rev.*

3.2. The environmental department is responsible for keeping facilitywide inventories of chemicals by type and quantity. The environmental department must
also compare this data to emissions and discharges to determine if leaks have occurred.

4. Procedure

4.1. Identify Responsible Individual

A person who is responsible for all chemical tracking at the site will be identified. Many people may be involved; however, one individual should coordinate the overall effort for the entire organization.

4.2. Inventory

Inventory the chemicals planned and on-site and enter this information into
a software system. For example, it should be indicated whether the chemical is
planned or on-site. The quantity of the chemical in use, storage and being discharged needs to be recorded on an ongoing basis. The discharged amount
would be in accordance with permits and regulations and include that to sewers,
water bodies, air and landfill. The amounts presently being recycled, reused or
sold should also be added into the inventory and so designated. Many of these
figures could be obtained from manifests, Superfund Amendments and Reauthorization Act (SARA) documents and other reports.

4.3. Other Information

Additional information should be entered into the software system, however,
this may require some research into material safety data sheets or calls to the
manufacturer. For example, the composition of the chemical along with Chemical Abstract System (CAS) numbers should be added. Hazard rating information should also be added.

4.4. Ongoing Review and Entry of Information

There will be on-going review of invoices, shipping documents, manifests,

TITLE:	PROCEDURE NO:	Page _____ of _____
Handling Empty Chemical Containers		
Company Name:	**Date:**	*Rev.*

plans and other information, which is added to the software system. The system should track, balance and account for, all chemicals. This helps to ensure that some have not leaked or been spilled from their containers.

4.5. Enforcement of Chemical Purchase Procedures

As all the numbers are obtained, there many be cases where it is found some individuals are not following the chemical purchase procedures. When this happens they will be reminded of the procedure.

4.6. When the Numbers Don't Add Up

If the mass balance shows that there are significant volumes of chemicals unaccounted for some additional research will be needed. If it is not just an error in calculation, it might be a leaking storage tank or a spill.

5. Related Documentation

5.1. Chemical Purchase Orders

5.2. Chemical Inventories

5.3. Tank Volume Reading Records

5.4. Hazardous Waste Manifests

TITLE:	PROCEDURE NO:	Page _____ of _____
Site Closure		
Company Name:	Date:	*Rev.*

Company Name: _____

Site: _____

Site Closure

This is a "controlled" document.
Routine distribution is restricted to the approved
distribution in _____. All other persons
in possession of this document have uncontrolled copies
and should call document control for revision level status.

Approved by _____

Approved Date _____

TITLE:	PROCEDURE NO:	Page _____ of _____
Site Closure		
Company Name:	Date:	*Rev.*

Site Closure Procedure

An organization must properly close down their operation in an environmentally conscious manner. This is required by law and a full disclosure must be made to the buyer of the property. Some of the procedural steps should include:

1. Purpose

1.1. To insure that an operation leaves the location in the best shape possible in terms of the environment.

2. Scope

2.1. This procedure covers the abandonment period for all types of organizations, whether they are heavy industrial or not.

2.2. This procedure covers operations where the land and buildings were owned and leased.

3. Responsibilities

3.1. It is the responsibility of the facility department to inventory remaining chemicals, equipment and all other facilities.

3.2. It is the responsibility of the environmental department to inventory hazardous and nonhazardous waste, close out environmental permits and prepare the closure plans specified in this procedure.

TITLE:	PROCEDURE NO:	Page _____ of _____
Site Closure		
Company Name:	Date:	*Rev.*

4. Procedure

4.1. Inventory

An inventory should be made to assess materials and areas requiring attention. For example, on-site raw materials, wastes, chemicals, fuels, production equipment, waste treatment equipment and electrical transformers should be itemized.

4.2. Phase I Environmental Assessment

A Phase I environmental assessment may be necessary and would be prepared by a consultant. This would especially apply to all manufacturing, research and distribution facilities with a history of environmental problems or where hazardous materials or hazardous waste have been stored.

4.3. Regulatory Requirements

An identification of local, state and federal requirements applicable to site closure should be made. For example an EPA ID number may have to be closed out.

4.4. Closure Plan

Depending on the site, results from the Phase I environmental assessment and the regulatory requirements identified, a written closure plan may be necessary. This plan would provide details concerning all of the actions presented below

4.5. Disposition of Remaining Materials

All remaining raw materials, chemicals, fuels and debris should be removed. The inventory prepared earlier should be checked off as these materials are being removed. If any of this can be sold or recycled, it would be better for the environment and the organization's liabilities.

TITLE:	PROCEDURE NO:	Page _____ of _____
Site Closure		
Company Name:	Date:	*Rev.*

4.6. Cleaning of Equipment

Production equipment, storage facilities, waste treatment and all other equipment that encountered chemicals or hazardous wastes must be thoroughly cleaned. This would include piping and ducts. Disposition of the cleaning wastes must be according to the regulations and will probably be considered hazardous waste. Tests should be done on the cleaned equipment to make sure that chemicals are at an acceptable level. This level of acceptance would depend on the landfill or buyers' requirements.

4.7. Special Hazards

Special procedures and handling would be necessary if asbestos, PCBs, on-site disposal or other special hazards are present. Numerous regulations specify how these substances are to be handled.

4.8. Notification

The applicable government authorities must be notified of site closure. This may also include submittal of a closure plan and ending an EPA ID number. The notification should be done in writing.

4.9. Permit Review

All environmental permits should be reviewed in terms of closure and appropriate action taken. A permit might require that an area or pond be reclaimed, for example. Equipment removal and decontamination are also required in some cases.

4.10. Photos

Photographs should be taken to document the condition of the interior and exterior of the site after completion of the closure activities. This is especially important if the next owner alleges that certain structures were left behind.

TITLE:	PROCEDURE NO:	Page _____ of _____
Site Closure		
Company Name:	Date:	*Rev.*

4.11. Final Site Inspection

A thorough inspection of the facility after all closure activities are completed is necessary. This will help ensure that closure has taken place in accordance with the procedures outlined above.

4.12. Closure Report

A report documenting the findings of the final site inspection should be made and filed as specified below. All observations should be recorded in case disputes arise in the future.

5. Related Documentation

Environmental records should be collected and transferred to a secure location. Many of the records, such as manifests, must be retained for a minimum of three years. Other records which should be saved include:

5.1. Closure Inventory

5.2. Closure Plan

5.3. Closure Report

5.4. Permits

5.5. Phase I Environmental Assessment

5.6. Purchase Agreements

5.7. Cleanup Testing

5.8. Photographs

5.9. Hazardous Waste Training Records

TITLE:	PROCEDURE NO:	Page ____ of ____
Site Closure		
Company Name:	**Date:**	*Rev.*

5.10. Bills of Lading, Shipping Papers and Consignment Notes

5.11. Chemical Purchase Records

5.12. Waste Testing Results

5.13. Reports Required by Permits and Any Other Environmental Record that Seems Appropriate at the Time of Closure

TITLE:	PROCEDURE NO:	Page _____ of _____
Procedure for External Communication		
Company Name:	Date:	*Rev.*

Company: _____

Site: _____

Procedure for External Communication

This is a "controlled" document.
Routine distribution is restricted to the approved
distribution in _____. All other persons
in possession of this document have uncontrolled copies
and should call document control for revision level status.

Approved by _____

Approved Date _____

Confidential

TITLE:	PROCEDURE NO:	Page _____ of _____
Procedure for External Communication		
Company Name:	Date:	*Rev.*

Procedure for External Communication

1. Purpose

1.1. To receive, consider and respond to appropriate communication from the public.

2.2. To provide timely, accurate and meaningful environmental communication to the public.

2. Scope

2.1. Communications will be received from all components of the public including customers, media, environmental groups, agencies, and so on.

3. Responsibilities

3.1. The public affairs department will receive all external communication and direct environmental questions and comments to the environmental department.

3.2. The environmental department will research and respond to external environmental communications, with the oversight of the public affairs department.

4. Procedure

4.1. Customer Communications

Most organizations get numerous customer requests for environmental information. Hopefully, when an organization has an ISO certification that covers environmental management, they will just have to show the certification. Presently, this is not the case. Now numerous requests for different types of environmental information require research and unique answers. A tracking system is needed so that all customer requests and responses are indexed by name, date

TITLE:	PROCEDURE NO:	Page ___ of ___
Procedure for External Communication		
Company Name:	Date:	*Rev.*

and subject. It is important that one group in the organization respond to all customer requests for environmental information. This will ensure consistency and accuracy.

The organization has an obligation to notify their customers of any chemical hazards present in the product. A material safety data sheet (MSDS) is the common avenue for doing this. The customer should be urged to reuse, recycle or, if all else fails, to dispose of the product in an environmentally conscious way.

Customer requests will pertain to many subjects such as the organization's use of CFCs, PBBs, lead, cadmium and recyclable packaging materials. The following list is a sample procedure for handling customer requests for environmental information:

4.1.1. Whoever receives the client inquiry should attach as much information as is readily available to the request and forward it to the environmental department which issues all responses.

4.1.2. The environmental department will determine if there is a regulatory or legal (contractual) requirement to provide the information.

4.1.3. If there is not a legal or regulatory requirement then the client coordinator will determine whether there is an important business reason to respond.

4.1.4. If the response is made, it must be accurate and consistent with responses made to other customers.

4.1.5. In some cases a response may require laboratory testing or extensive research. The customer should be notified that there will be a time delay and possible increase in cost of the product or service.

4.2. Media Communications

It is a good idea to have a well-established procedure for handling contacts by the media since these may be sensitive exchanges. It is in everyone's best interest that the information is reported in a factual and nonemotional way. A few suggestions follow:

TITLE:	PROCEDURE NO:	Page ____ of ____
Procedure for External Communication		
Company Name:	**Date:**	*Rev.*

4.2.1. Establish a site-specific procedure before a sensitive issue develops.

4.2.2. Whoever receives the call should transfer it to the public relations department within the organization.

4.2.3. Both the public relations individual and the technical individual should return the call. All answers should be brief and factual.

4.3. General Public Communications

When someone from the general public has contacted the organization, the comment should be documented. An analysis should next occur to determine whether the comment has merit. The determination and reasons should also be recorded and relayed to the individual.

5. Related Documentation

5.1. Record of All External Communication Received from the Public

5.2. Copies of All External Communications Issued to the Public

TITLE:	PROCEDURE NO:	Page _____ of _____
Data Collection and Handling Procedure		
Company Name:	Date:	*Rev.*

Company Name: _____

Site: _____

Data Collection and Handling Procedure

This is a "controlled" document.
Routine distribution is restricted to the approved
distribution in _____. All other persons
in possession of this document have uncontrolled copies
and should call document control for revision level status.

Approved by _____

Approved Date _____

Confidential

TITLE:	PROCEDURE NO:	Page _____ of _____
Data Collection and Handling Procedure		
Company Name:	Date:	*Rev.*

Data Collection and Handling Procedure

1. Purpose

1.1. To obtain the best information possible so that meaningful interpretations can be made.

2. Scope

2.1. Data should be collected from any source that is reliable and not too dated.

2.2. Data should be collected from the site itself as a top priority and then from off-site sources if they are meaningful.

3. Responsibilities

3.1. It is the responsibility of the environmental department to collect and interpret all data except routine process data.

3.2. Routine process data should be collected by the operators of the equipment.

4. Procedure

4.1. Sample Collection

The most cost-effective, safe and accurate method to collect the samples will be specified. For example, some samples will be either composite or grab. Composite samples are usually for lesser toxic substances while grab are commonly for hazardous samples. A grab sample may not be as accurate, however, it minimizes possible exposure time for the sampler.

TITLE:	PROCEDURE NO:	Page ____ of ____
Data Collection and Handling Procedure		
Company Name:	Date:	*Rev.*

4.2. Sample Containers

Place the samples in proper sample containers. Care is needed since certain samples can dissolve glass or plastic containers. Breakage during transportation is also a consideration.

4.3. Transport of the Samples

Usually the samples should be transported quickly and in containers that will not break. In some cases this must also be done with preservative or ice. For example, volatile organic compounds are harder to transport since they may release vapors.

4.4. Laboratory Analysis

The samples will be analyzed in a certified laboratory. Even if some analysis can be done using field instrumentation, usually a laboratory is used for greater accuracy. Certified laboratories must have detailed quality control (QC) systems.

4.5. Data Interpretation

The data will be analyzed in order to make sound decisions. Scientific methods, research methods and interpretative processes all basically take numbers and extract meaning from the numbers. It is not enough to say there are 5 ppm of lead. Is this within safe limits and does it comply with the regulations?

4.6. Monitoring Equipment Maintenance

All monitoring equipment will be maintained and routinely calibrated. This applies to whether the organization performs the monitoring or a consultant. Field sample equipment and field and lab analysis equipment all must be properly maintained. Maintenance activities should be recorded.

4.7. Management of Data

There are numerous data management systems available on the market today.

TITLE:	PROCEDURE NO:	Page _____ of _____
Data Collection and Handling Procedure		
Company Name:	**Date:**	*Rev.*

Many of these involve high powered software systems. For example, MSDS information can be loaded into a chemical data base and combined with chemical inventory information. This allows the environmental manager to track chemical usage and general information better. Recycling data, injury data, discharge data and most other environmental information take on new meaning if they are managed in this way.

4.8. Use of the Data

The data collected will be used to adjust objectives and targets and to improve processes and systems. This requires adjustment based upon the data. The data should also be recorded in order to show compliance. This will help if fines are assessed or litigation develop. The data, or a summary of it, should also be reported to senior management.

5. Related Documentation

5.1. Process Data

5.2. Monitoring Data of Emissions, Discharges and Receiving Medium

5.3. Permit Requirements Concerning Data

5.4. Risk Assessments

TITLE:	PROCEDURE NO:	Page _____ of _____
Disaster Recovery Plan Procedures		
Company Name:	Date:	*Rev.*

Company: _____

Site: _____

Disaster Recovery Plan Procedures

This is a "controlled" document.

Routine distribution is restricted to the approved

distribution in _____. All other persons

in possession of this document have uncontrolled copies

and should call document control for revision level status.

Approved by _____

Approved Date _____

Confidential

TITLE:	PROCEDURE NO:	Page ____ of ____
Disaster Recovery Plan Procedures		
Company Name:	Date:	*Rev.*

Disaster Recovery Plan Procedures

1. Purpose

1.1. To return the business back into operation in a safe, quick, cost effective and environmentally sensitive way.

2. Scope

2.1. This procedure applies to the period of time immediately after the emergency response team has controlled the immediate crisis.

2.2. This procedure would apply to any disaster including chemical spills, fires, severe weather, and so on.

3. Responsibilities

3.1. The emergency response team is responsible for dealing with immediate threats to employees and the environment.

3.2. The disaster recovery team is responsible for dealing with return of the business into operation, protection of the public, and environment.

4. Procedure

4.1. Up-Front Activities

    4.1.1. Assembling the Emergency Response Team (ERT)/Disaster Recovery Team

    In order to prepare a meaningful disaster recovery plan, a team will be assembled. After they have prepared the plan, they will meet thereafter at least once per year or in the event of a disaster. The members would include the ERT plus

TITLE:	PROCEDURE NO:	Page _____ of _____
Disaster Recovery Plan Procedures		
Company Name:	Date:	*Rev.*

MIS, production, materials, facilities, environmental/health/safety, security, sales, engineering and quality.

4.1.2. Identification of Site Resources

An inventory of critical operations and resources will be made. In case the site is destroyed in part or in total, this inventory will suggest what needs to be quickly replaced. The inventory should include a description of people, files, products and raw materials and can probably be assembled from current documents.

4.1.3. Impact Assessment

An assessment of what might happen to each of the critical types of resources identified above will be made for possible disaster scenarios. This will help determine where backup is needed. The assessment can be in a variety of formats; however, Table 13-1 presents one possibility. The ranking shown is based on probability. For the generic site presented in Table 13-1, a chemical spill has the greatest potential of occurring.

4.1.4. Impact Minimization Strategies

Based on items 4.1.2. and 4.1.3., an impact minimization strategy will next be prepared for the resources identified which are critical and have a high likelihood of being impacted or destroyed. For example, this may include more training, backup files and operations at other locations, secondary or tertiary containment around some chemicals and wastes, upgrading the ER plan, earthquake bracing and additional fire suppression systems (sprinklers, hose reels, fire extinguishers).

4.1.5. Recovery Strategy

The preceding impact minimization steps would be done before the disaster. If they are done well there will be fewer recovery actions necessary after the disaster. Usually some recovery actions will still be necessary. Also certain impacts cannot be minimized or avoided if a severe disaster occurs. In these situations all the organization can do is have some recovery strategies ready and then pick up the pieces the best they can, such as after a wide-area, large earthquake or war.

TITLE:	PROCEDURE NO:	Page ____ of ____
Disaster Recovery Plan Procedures		
Company Name:	Date:	*Rev.*

4.1.6. Phone Numbers and Contacts

There should be even more emergency numbers in the disaster recovery plan than in the emergency plan. Emergency phone numbers will be documented in the disaster recovery plan. In addition to the ones already identified in the ER plan, the following should be included:

- Property owner
- Civil defense
- Upper management

4.1.7. Routine Inspections

The organization's status, resources and supplies in terms of disaster recovery are going to vary greatly from day to day. Therefore the company resources, disaster recovery supplies and plan need to be inspected frequently. As the resources change, the disaster recovery plan should be upgraded.

4.1.8. Disaster Recovery Headquarters

An off-site disaster headquarters location will be established. This could be in another company facility as long as it is located at least a few miles from the site in question. It is not a good idea to put the disaster recovery headquarters in the corporate headquarters, as most organizations do.

4.1.9. Preventative Maintenance

If production and environmental control equipment is maintained it will make it through a disaster will less impact to the operation and environment. Poorly maintained equipment, tanks and piping will be the first things to go when a disaster strikes.

4.1.10. Backup of Computer Files and Systems

Critical data, especially on hard drives, will be backed up and stored off-site on a weekly basis. Also major operational software systems essential to production should be able to function at other locations, in addition to the ones which may be destroyed.

TITLE:	PROCEDURE NO:	Page _____ of _____
Disaster Recovery Plan Procedures		
Company Name:	Date:	*Rev.*

4.1.11. Back-up of Paper Files

Paper-type files which are critical to the operation should be copied and filed at a backup location. An alternative would be to store backup on floppies or microfiche in a fire-resistant safe.

4.1.12. Communications

Phone systems notoriously go down during a disaster and severely cripple the operations recovery. Backup capability is advised. This usually means cellular phones.

4.1.13. Employee Supplies

Some supplies should be purchased, before a disaster, for the health and safety of the employees. These might include water, blankets, flashlights, tools and food. There is usually much discussion about the proper amount of supplies to have on hand, especially since some, like water and food, have a limited shelf life. Having enough water and food for 75 percent of the population for three days is a minimum.

4.1.14. Environmental Protection Supplies

In addition to the spill cleanup supplies mentioned in the emergency response plan, it would be a good idea to have some environmental protection supplies oriented toward minimizing environmental impacts during a disaster. For example, depending on the operation, back-up drums and pumps in case large tanks are destroyed during a disaster are recommended.

A Risk Management and Prevention Plan (RMPP) is required for certain operations that store large amounts of toxic or extremely hazardous materials, such as chlorine and sulfuric acid. This type of assessment and plan is oriented toward protection of the off-site public and environment and may even require items like backup drums and pumps. If an organization already has a RMPP or similar plan it should be incorporated into the disaster recovery plan.

TITLE:	PROCEDURE NO:	Page _____ of _____
Disaster Recovery Plan Procedures		
Company Name:	Date:	*Rev.*

4.1.15. Facility Drawings

All important facility drawings will be assembled and stored at the off-site disaster headquarters. If this becomes too cost prohibitive, then at least the drawings showing utilities, tanks, piping and chemical and hazardous waste storage should be duplicated and stored off-site.

4.1.16. Copying and Distribution of the Plan

For obvious reasons it is important to prepare and distribute the plan *before* a disaster occurs. This entire disaster recovery section could serve as a generic outline for the plan and then site-specific information would be added. Once the plan is completed, it will be given to the ERT, the disaster recovery team, the security command post, site environmental/health/safety representatives, security and upper management. The plan should be updated at least once per year or earlier if major changes have occurred.

4.2. During and Immediately Following a Disaster

After the ERT has the immediate crisis under control, as specified in the ER plan, the following activities should occur. These would be considered the recovery actions specified in the preceding Recovery Strategy step (step 4.1.5.).

4.2.1. Convene ERT/Disaster Recovery Team

The ERT will already be assembled and have addressed the immediate health and safety issues and now transition into a disaster recovery team (DRT). Additional members will be added at this time and include MIS, production, materials, operations and finance.

4.2.2. Inspection of the Area

The team would do an initial audit for safety hazards and, if any are found, notify employees to stay out. Additional inspections of the area would also be made later and a business damage assessment completed. The subsequent inspections should include photos, dollars required to come back into operation and recommendations.

TITLE:	PROCEDURE NO:	Page _____ of _____
Disaster Recovery Plan Procedures		
Company Name:	Date:	*Rev.*

4.2.3 Employee Needs

Even if the immediate safety needs have already been addressed, the longer-term needs should now be considered. These may include providing information to families or help in locating families. Employees may have other needs during and after significant disasters. For example, if a large earthquake has occurred, food, water, blankets and temporary shelter may also be needed by the employees.

4.2.4. Insurance Companies

The property insurance representative should be called and visit the site immediately. This individual may recommend companies that can assist in the recovery. They should be called early, before repairs start, so that maximum coverage will be provided. Sometimes this does not happen if urgent repairs have to start quickly.

4.2.5. Skills Bank

All employees who are able should report to an office that has been set up to match their skills to disaster recovery jobs they can perform safely. This not only helps the organization but also helps some of the employees deal with the disaster better since they feel productive.

4.2.6. Reestablish Utilities

During the disaster some of the utilities may have shut down by accident or by design. The DRT should work with the gas, electric, water and sewer utilities to restore service. If chemicals are present it would be best to get the electricity on first so that ventilation systems can start clearing vapors from the area.

4.2.7. Reestablish Communications

Close work with the telephone company may be needed to reestablish phone lines. In the interim it may be necessary to use cellular phones, radios or other forms of communication. Once the main telephone system is back in operation, a hotline should be established to answer employee and public questions and concerns.

TITLE:	PROCEDURE NO:	Page _____ of _____
Disaster Recovery Plan Procedures		
Company Name:	Date:	*Rev.*

4.2.8. Facility Repair

The DRT should help relocate the operation, if needed, to an alternate space and/or start restoring the damaged facility. Significant damage to the facility may necessitate a temporary location. The location could be in an undamaged area or completely off-site if necessary. The first system that should probably be fixed is the fire protection system followed by the security command post operation.

If a fire has occurred, the fire suppression system probably left everything wet. Moist equipment and supplies should be removed or dried promptly to prevent rust, mildew and health problems. Heat-damaged equipment and insulation should be repaired or replaced. Some smoke damage may also have occurred which could lead to product contamination or corrosion if it is not removed.

4.2.9. Inspection and Repair of Chemical and Hazardous Waste Storage Structures

Assuming chemical and hazardous waste leaks and spills have been addressed during earlier ERT activities, it is now time to verify that the chemical and waste systems are all sound and not impacting the environment in a less obvious way. If a potential problem is noted, it should be checked out and if necessary corrected immediately.

4.2.10. Reestablish MIS/Computer Systems

Most operations are totally dependent upon their computer systems. These must be repaired or relocated quickly. The backup system that supports the customers should be one of the first to be reestablished.

4.2.11. Replacement of Critical Files

Any essential files which may have been destroyed should be re-created by off-site files. This is especially important concerning certain customer and personnel files. If floppies or microfiche were made, this task would not be as involved.

TITLE:	PROCEDURE NO:	Page ____ of ____
Disaster Recovery Plan Procedures		
Company Name:	Date:	*Rev.*

4.2.12. Reestablish Financial and Human Resource Systems

It may be necessary to shift certain financial and HR systems, such as compensation and benefit administration to another site for a period of time. The logical location is where backup records are kept. Either hard copies, microfiche, or electronic storage would be acceptable. Key financial records should be copied and stored off-site every 24 hours so that they will be available now to help the operation reestablish itself.

4.2.13. Dealing with the Media

All contacts from the media will be directed to the site's public relations manager. No other employees should make statements. Hopefully any media releases will help gain support and assistance for those impacted by the disaster. Sometimes, however, the media coverage just adds chaos and emotional distress.

5. Related Documentation

5.1. Emergency Response Plan

5.2. Inventory of Damaged Facilities

5.3. Spill Report Records

5.4. Injury Report Records

TITLE:	PROCEDURE NO:	Page ____ of ____
Procedure for Minimizing Discharges to Air, Water Bodies and Sewers		
Company Name:	**Date:**	*Rev.*

Company: _____

Site: _____

Procedure for Minimizing Discharges to Air, Water Bodies and Sewers

This is a "controlled" document.
Routine distribution is restricted to the approved
distribution in _____. All other persons
in possession of this document have uncontrolled copies
and should call document control for revision level status.

Approved by _____

Approved Date _____

Confidential

TITLE:	PROCEDURE NO:	Page ____ of ____
Procedure for Minimizing Discharges to Air, Water Bodies and Sewers		
Company Name:	Date:	*Rev.*

Procedure for Minimizing Discharges to Air, Water Bodies and Sewers

1. Purpose

1.1. To measure the quantity and quality of pollutants coming from processes and then design emission controls and other methods to minimize impacts.

2. Scope

2.1. This procedure applies to point and nonpoint sources of pollutants.

3. Responsibilities

3.1. It is the responsibility of the environmental department to set up monitoring programs for discharges, to oversee data collection and to suggest discharge minimization opportunities.

3.2. The operators of the equipment creating the discharge are responsible for collecting routine samples, maintenance of equipment and suggesting minimization ideas.

3.3. It is the responsibility of the engineering or production departments to implement cost-effective discharge minimization changes.

4. Procedure

4.1. Baseline

Base-line air or water quality measurements will be taken. This will provide valuable information for future reference. It will show progress made by the organization and might also differentiate between present pollution and past operation pollution.

TITLE:	PROCEDURE NO:	Page _____ of _____
Procedure for Minimizing Discharges to Air, Water Bodies and Sewers		
Company Name:	**Date:**	*Rev.*

4.2. Emission Estimates

Before detailed design occurs, it is important to make estimates of the proposed emissions. Even if the estimates are rough, they will begin to allow discussions concerning possible control equipment and systems.

4.3. Regulatory Requirements

Check the applicable regulations and then contact the regulatory authority and obtain a permit to construct. This should be done very early in the process since expensive controls may be required.

4.4. Impact Minimization

It is important to reduce as many potential emissions during the design phase as possible through controls and waste minimization. This may cost some up-front capital, however, there should be a cost savings over the life of the operation.

4.5. Installation

During the installation of the process and control equipment attention will be directed toward regulatory requirements. Permit stipulations must be followed very closely. If modifications are needed, and most complex installations result in some modification, the significant changes must be run by the regulatory authority. The agency should be notified, especially if new discharge points are added or if permit limits will be exceeded.

4.6. Training

Operators should be trained in terms of the process and environmental considerations. This training is necessary in terms of employee safety as well as environmental protection. Training will be both initial and with annual refreshers depending on the class.

TITLE:	PROCEDURE NO:	Page _____ of _____
Procedure for Minimizing Discharges to Air, Water Bodies and Sewers		
Company Name:	Date:	*Rev.*

4.7. Permit to Operate

Once the installation and training has been completed, the company should obtain a permit to operate and start operations. In some locations a separate type of permit is required. The regulatory agency will probably need to come out for inspections, usually after the process has operated for a period of time.

4.8. Sampling/Monitoring

Many regulations require the sampling of actual discharge or emissions from the process and verification of regulatory compliance. At a minimum, the variables specified in the permit will be assessed. Ongoing monitoring will also ensure compliance with company policies and continuous improvement of the system.

4.9. Upgrade Controls

Depending on the actual data obtained, it may be necessary to improve controls. This may also be just a matter of adjusting the existing control equipment. Sometimes, however, major modifications will be needed to allow the controls to remove pollutants to regulated levels.

4.10. Maintenance

Both the process and control equipment will be carefully maintained.. This is very important since many problems are related to lack of proper maintenance.

4.11. Logs

It is good management to keep logs of fuel and chemical usage. Some permits may require logs, and, even if they do not, this type of information is needed for mass balance tracking.

TITLE:	PROCEDURE NO:	Page _____ of _____
Procedure for Minimizing Discharges to Air, Water Bodies and Sewers		
Company Name:	**Date:**	*Rev.*

4.12. Audits

Routine audits of the operation will be made and adjustment carried out to continually reduce emissions, chemicals and wastes. The audits would be done in response to special problems and scheduled or routine audits for regulatory compliance.

4.13. Special Problems

Unauthorized discharges, spills and other problems must be reported immediately to company management and to the regulatory authority. In certain cases, forms will have to be completed.

5. Related Documentation

5.1. Discharge Data

5.2. Maintenance Records

5.3. Waste Minimization Project Cost Benefit Analysis

TITLE:	PROCEDURE NO:	Page _____ of _____
Emergency Procedures		
Company Name:	Date:	*Rev.*

Company: _____

Site: _____

Emergency Procedures

This is a "controlled" document.
Routine distribution is restricted to the approved
distribution in _____. All other persons
in possession of this document have uncontrolled copies
and should call document control for revision level status.

Approved by _____

Approved Date _____

Confidential

TITLE:	PROCEDURE NO:	Page _____ of _____
Emergency Procedures		
Company Name:	Date:	*Rev.*

Emergency Response Plan Procedures

1. Purpose

1.1. To minimize the impact of emergency events on employees, the public and the environment.

2. Scope

2.1. This procedure covers the first hours of an emergency.

2.2. The procedure is oriented toward minimizing initial impact to employees, the public and the environment.

3. Responsibilities

3.1. It is the responsibility of the emergency response team (ERT) to prepare and implement the emergency response plan.

3.2. It is the responsibility of all employees to understand their evacuation routes and to follow the instructions of the ERT to the letter.

4. Procedure

The emergency plan will be prepared by the site environmental, health and safety individuals who have knowledge of site conditions and local regulations. Security, facilities, legal and human resources should also contribute to the plan or at least complete a review. The emergency response team itself must also be involved in the preparation of the plan or at least in the continual upgrade of the plan over time.

The plan will be amended when important components become outdated or business and regulatory changes dictate. Whenever changes are made, a revision

TITLE:	PROCEDURE NO:	Page _____ of _____
Emergency Procedures		
Company Name:	Date:	*Rev.*

date and number must be noted. In addition to the revision date each plan/
binder should be numbered for tracking purposes. All changed pages should be
sent to all plan holders with a controlled copy. When a plan holder receives an
update they must sign and return a transfer memo to the department coordinating the plan.

Any plan issued without a number would be considered an uncontrolled
copy. It is the responsibility of the holder of any uncontrolled copy to request
updates. Any copies made of uncontrolled plans or portions of plans should be
clearly marked "Uncontrolled Copy."

Many different people in the organization should be given a copy of the plan
and everyone should have easy access, if not their own copy. At least one copy
will be in every building; it is usually located at the reception desk, guard station
or wall-mounted box near the front door. In addition, the following individuals
will have their own "controlled copy": incident commander, each ER team
member, safety committee chair, site environmental, health, and safety representative, local fire department and local hospital.

4.1. Up-Front Activities and Components Which Should Be Prepared before an Emergency

All emergency response plans are unique to a particular operation. They must
be specific if they are going to be useful in an emergency. There are a few key
up-front elements of most emergency response plans and they are as follows:

4.1.1. Forming the ER Team

The team will be composed of employees who have knowledge of can be
trained in responding to emergencies such as fires, explosions, chemical spills
and so forth. Designate an adequate number of employees to serve as emergency
response team (ERT) members. There must be teams for all shifts and team
members need to have alternates. Each team needs a team leader.

Most organizations will require each functional department to designate one
person to serve on the ERT. If this does not generate an adequate number of

TITLE:	PROCEDURE NO:	Page ____ of ____
Emergency Procedures		
Company Name:	Date:	*Rev.*

ERTs then they could be assigned by building.   Due to the great amount of training required, it is most cost-effective to require the ERT member to serve at least two years and longer if they desire.

4.1.2. Personnel Protective Equipment

Personnel protective equipment (PPE) will be stationed in strategic locations for the ERT. Depending on the chemicals in the facility, this may include self contained breathing apparatus (SCBA), chemical suits, gloves, boots, respirators, cartridges and duct tape. Depending on the PPE used, it may be necessary to do fit tests, such as for respirators, before an emergency occurs.

4.1.3. Chemical Spill Cleanup Supplies

Before an emergency occurs it is necessary to purchase spill cleanup supplies and station the supplies in high-risk areas. The type of chemical again dictates the supplies needed; however, these commonly include absorbent pillows, pads or booms, acid and base neutralizers, pH paper, disposal bags or drums, hazardous waste labels, barricade tape, storm drain covers, tool kits, maps, brooms, shovels and squeegees.

4.1.4. Training

Training of the ERT is required in how to handle various situations such as spills, fires, injuries, earthquakes and extreme weather problems. First aid and CPR training are the first instructions the ERTs should get. These will be followed by hazardous waste operations and emergency response (HAZWOPER) first responder training which is required by OSHA for all ERTs. Depending on the chemicals in the facility the training must last from 24 to 40 hours. Subjects covered include respiratory protection, toxicology, incident command systems, spill cleanup procedures, handling of drum emergencies, hazard classification, MSDS understanding, hazard identification and assessment, PPE, monitoring equipment, first aid, fire control, DOTs Emergency Response Guide, decontamination and numerous other general and site-specific topics.

It is very important that management support the ERT training. The supervisor must allow time for training and insist that their employee is properly

TITLE:	PROCEDURE NO:	Page _____ of _____
Emergency Procedures		
Company Name:	Date:	*Rev.*

trained concerning the ERT functions. Without this type of strong support, the ERT may not be given the time for the training. This could result in an injury of the ERT and possibly of other employees. The site environmental, health and safety representative and ERT incident commander should also encourage and document that the necessary training is occurring.

4.1.5. ERT Drills

The ERT will go through drills to practice the skills they learned during training. For example, they will have mock spills to ensure they are following the right procedures. The drills should occur at least every other month, with a discussion of the success and problems encountered. The drill could occur during a scheduled monthly meeting of the ERT and occasionally unannounced.

4.1.6. Physicals

All members of the ERT will undergo physicals and respirator fit tests. Medical examinations must be required to minimize the chance that an ERT cannot perform the function because of some physical limitation. A doctor should certify that each ERT is physically fit to participate in ERT activities

4.1.7. ERT Communication

Pagers or some other form of emergency communication equipment must be provided to the ERT. The ERT members must have a pager, radio or cellular phone so that they can be summoned to the scene quickly. The pager numbers need to be given to the security command post, reception desks, internal operators, site environmental/health and safety representatives and numerous other people. It is also a good idea to provide some type of emergency communication device to the site environmental/health/safety representative, security guard and site nurse.

4.1.8. Emergency Response Plan

The entire emergency response plan needs to be prepared before an emergency occurs. The plan should be updated at least once a year or sooner if significant changes have occurred.

TITLE:	PROCEDURE NO:	Page _____ of _____
Emergency Procedures		
Company Name:	**Date:**	*Rev.*

4.1.9. Availability of the Team

At a minimum the ERT will be available during operating hours of the facility. For continuous operation this would mean that the ERT should be on-site 24 hours a day, 7 days a week. Obviously there would need to be several teams to cover all the shifts. Even if the facility does not operate continuously the ERT should be on call in case of a leaking tank or other problem that occurs during off-hours.

4.1.10. Designated Internal Emergency Phone Number

A dedicated site emergency phone number will be set up that can be used from any in-house telephone. It is best if the phone number is a simple number to remember, such as 2222. When someone calls they need to be immediately connected to the security command post, who will dispatch the ERT and/or proper community agency (fire, police ambulance, etc.). Callers should be instructed to stay on the phone long enough, if it is safe to do so, to answer questions such as those shown in Figure 13-1. If the caller is physically or psychologically unable to answer many questions, than the questions should concentrate only on their location.

4.1.11. External Emergency Phone Numbers

Phone numbers and guidelines will be provided to call the police, fire department and ambulance. The guidelines are very important since there are many gray area situations when the caller will be unsure whether it is really an emergency or not. For example, the following would be called immediately by the security command post:

- 911 or ambulance – individual(s) in life threatening condition
- Fire Department – actual fire

All other calls would be made by the ERT Incident Coordinator in conjunction with the site environmental, health and safety representative. The calls would be made after a quick determination of applicability and seriousness is made and may include (not arranged in order of priority):

TITLE:	PROCEDURE NO:	Page _____ of _____
Emergency Procedures		
Company Name:	Date:	*Rev.*

- State office of emergency services – most emergencies
- Waste water treatment plant – spills into the sewer
- Local and state water quality office – spills into streams, lakes, drainages
- Local and state environmental office – spills over the reportable quantity
- Fish and game office – spills which may impact wildlife
- Air pollution office – releases or spills that become airborne
- Local gas and electric company – gas leaks and power failures
- Highway patrol – emergencies on roadways (spills, etc.)
- Police department – local emergencies
- Hospital – medical emergencies
- Local clinic or on-site nurse – medical emergencies
- Fire department – fires, explosions, spills
- Local contract chemical/hazardous waste cleanup service – spills
- Chemtrec (800-424-9300) – chemical spills
- Company public relations office – any emergency
- Corporate offices of environmental, health and safety – any emergency

4.1.12. Evacuation Maps

Up-to-date evacuation maps will be prepared and posted in numerous locations in the facilities. These maps show the closest exit, backup exits, location of the ER plan, security/reception desk, fire extinguishers, eye wash, emergency showers, spill supplies, first aid supplies and other key elements. Employees should be told to memorize the primary route identified for them and a backup route in case the primary exit is blocked or involved in the emergency.

4.1.13. Public Address System

Some type of communication system will be ready in case of any emergency. The best would be a paging system or emergency alarms, with battery backup that can be heard everywhere in the facility. If this is not possible, then hand-held megaphones or some other public address system should be in place. It is very important that whatever system is selected must be heard in every area of the facility where employees might frequent. This includes remote areas, bathrooms, break areas and noisy areas. The public address system should be tested monthly to verify that it is operating properly.

TITLE:	PROCEDURE NO:	Page _____ of _____
Emergency Procedures		
Company Name:	Date:	*Rev.*

4.1.14. Outside Assembly Points

Several predesignated outside assembly points will be marked and employees instructed where to assemble in the event of an emergency. The supervisors should be told that this is where they take role once the evacuation has occurred. To do this efficiently the supervisor must be able to quickly tell who is on which shift, vacation and sick leave.

4.1.15. Other Emergency Equipment

In addition to the spill cleanup supplies, radios and PPE already mentioned, which are the most important, there is additional emergency equipment which will be obtained as well. Safety showers, eye wash stations, fire extinguishers, first aid supplies, bloodborne pathogen kits, stretchers, backboards, oxygen and decontamination equipment are examples of other useful equipment.

4.1.16. Practice Evacuations

At least once a year the entire employee population will have an evacuation drill. If a drill is too disruptive to production because of clearing out the entire facility at one time, then each department could do a separate drill. Also, if a few employees have to remain behind to maintain a critical process, they should be walked through their own drill after their shift. If at all possible, it is best to have the entire facility participate together in one drill, as it would in a real emergency.

During the drill and real emergencies, handicapped employees must be assisted from the area. The supervisor should assign two employees to assist each handicapped individual during the evacuation. This designation should be made well in advance of drills and emergencies. The individuals assigned must be physically able to carry out the handicapped employees, if needed.

Based upon notes taken by the ERT during the drill, a report should be given to all supervisors concerning success of the drill. Suggestions should be made to decrease evacuation time and increase overall control. Any special problems encountered should be noted along with corrective actions. Once the supervisors get the report, they should go over it with all their employees.

TITLE:	PROCEDURE NO:	Page ____ of ____
Emergency Procedures		
Company Name:	Date:	*Rev.*

4.2. Activities During An Emergency

Once all of the above is in place, the ER team should be prepared and able to respond quickly, efficiently and safely.   There are an infinite number of possible actions that could occur in an actual emergency. It would be impossible to list all appropriate actions since what is appropriate in one situation may not be advisable in another situation. The following are therefore only examples. An attempt was made to present these suggestions roughly in the order that they commonly occur. In certain situations, however, it may be necessary to reorder the following steps or do some in parallel.

4.2.1. Notification

The ERT is notified of an emergency by the security command center or some other source and the ERT assembles at the site of the emergency. The notification may be made by individual pagers, radios or a public address system. The public address system should be a last resort as this may cause employees to panic and at a time when the ERT would not yet be assembled to help manage the panic situation.

4.2.2. Evacuation

The ERT would clear employees out of the affected area if there is an immediate threat to human life. The decision to evacuate employees should be made by the ERT incident commander, with input from as many knowledgeable individuals as time will allow, such as the area supervisor. Employees should be told to move out of the area in an orderly manner through designated routes identified on the evacuation maps.

Once the alarm or announcement has been made, the ERT and supervisors should ensure that the evacuation is proceeding smoothly. For example, employees should not be panicking, not using elevator, and not picking up personal belongings. As the supervisor exits the area with their employees they should make a quick check of rest rooms or other areas where an employee might still remain.

TITLE:	PROCEDURE NO:	Page _____ of _____
Emergency Procedures		
Company Name:	Date:	*Rev.*

4.2.3. Accounting For Employees at the Assembly Points

It is the supervisor's responsibility to account for all their employees at the assembly point, considering those out on medical leave and vacation. If any employees are missing, the ERT incident commander must be notified immediately of the name and last known location. Employees must be told not to try to re-enter the area until the all-clear signal is given by the ERT incident commander.

4.2.4. Assessment of the Emergency

The ERT will put on PPE and inspect the area to make sure all employees are out and make an assessment of the emergency situation. The buddy system must be used during this assessment. The substance spilled or other cause of the emergency should be clearly identified by looking at labels, using hand-held meters or other methods.

4.2.5. Remove Injured Employees

If injured employees are found, they should be carefully moved out of the area of concern only by the ERT who must be wearing proper PPE. Depending on the injury it may be necessary to wait until an ambulance arrives.

4.2.6. Initial Calls to Outside Resources and Agencies

If immediate assistance is needed the ERT incident commander will instruct whom to call from the preceding list. This may include the fire department, ambulance and other emergency type agencies.

4.2.7. Shutdown of Certain Utilities and Services

During an emergency it may be necessary to shut down gas, electricity, water, sprinkler systems and other services. The incident commander of the ERT will make this decision with input from others such as the facilities department. Care must be taken to not shut down too much as this may hamper resolution of the emergency and cause serious disruption to the business.

TITLE:	PROCEDURE NO:	Page _____ of _____
Emergency Procedures		
Company Name:	Date:	*Rev.*

4.2.8. Erect Barricades

Barricades or barriers establish a zone of isolation that should prohibit entry by everyone except the ERT. The zone of isolation should consider that some vapors may travel great distances very fast. If this is the case, a considerable distance down-wind may have to be evacuated. The police and other agencies would definitely have to be involved in this type of evacuation if it involves more then just the employees of the organization in question.

4.2.9. Stop the Source of the Spill

The source of the spill should be stopped if it can be done safely. This is commonly done by plugging a hole or uprighting a drum, for example. Proper PPE and the buddy system should always be used, no matter how small the source or leak.

4.2.10. Relaying Information to Employees

The supervisors should relay factual information to their employees in an attempt to relieve their anxiety. Depending upon the seriousness of the emergency, some or all of the employees may be allowed to go home for the remainder of their shift. If this is done, names and destination of the released employees must be documented by the supervisor.

4.2.11. Clean Up the Spill

Again, if it can be done safely, the spill should be cleaned up. The ERT must be familiar with the substance and have the proper training, PPE and cleanup equipment before attempting to do this. Usually a well trained ERT can add absorbent or neutralizing materials to a spill if they know the spilled chemical. If there is any question, a professional spill cleanup vendor should be called. The cleaned-up material should be treated as hazardous waste.

4.2.12. Soil and Water

If a spill is headed for soil, a surface water body, storm drain or sewer, it should be stopped if it can be done safely. If the spill or other emergency has had an obvious or probable impact on soil or water, extra considerations must

TITLE:	PROCEDURE NO:	Page _____ of _____
Emergency Procedures		
Company Name:	Date:	*Rev.*

occur. An initial assessment of impact would be to determine whether more than the reportable quantity of the chemical has entered the soil or water. If so, it may be necessary to do a much more detailed assessment and a costly cleanup. All of these activities must be reported and closely coordinated with the applicable regulatory agencies.

4.2.13. Reporting

The spill should be reported to the applicable agencies if it has exceeded the reportable quantities (RQ) or if other serious considerations developed. This reporting must be done within prescribed time limits, depending on the agency or fines may be assessed.

4.2.14. Re-Entry of Building by Employees

The ERT incident commander will determine (with assistance of others) and announce when the building or area is safe to enter. No one else, without exception, should allow people back into the area.

4.2.15. Closing Meeting

The ERT, management representatives, environmental/health/safety representatives and agencies involved should have a meeting after the incident is over to discuss problems and corrective measures to minimize future occurrence. Certain results of the meeting should be relayed to the affected employee population to help relieve anxiety.

5. Related Documentation

5.1. Disaster Recovery Plan

5.2. Evacuation Maps

5.3. Inventory of Spill Cleanup Supplies

5.4. Spill Reports

5.5. Records of Evacuation Drills

TITLE:	PROCEDURE NO:	Page _____ of _____
Identification and Tracking of Financial Resources to Cover Environmental Management		
Company Name:	Date:	*Rev.*

Company: _____

Site: _____

Identification and Tracking of Financial Resources to Cover Environmental Management

This is a "controlled" document.
Routine distribution is restricted to the approved
distribution in _____. All other persons
in possession of this document have uncontrolled copies
and should call document control for revision level status.

Approved by _____

Approved Date _____

Confidential

TITLE:	PROCEDURE NO:	Page ___ of ___
Identification and Tracking of Financial Resources to Cover Environmental Management		
Company Name:	Date:	*Rev.*

Procedure for Identification and Tracking of Financial Resources to Cover Environmental Management

1. Purpose

To identify money needed to properly minimize environmental impacts of the organization to the greatest extent practical. Money will have to be spent to either proactively minimize environmental impacts or to clean up environmental problems once they occur. Obviously, proactive management is going to cost the organization considerably less over the long term than to perform a massive cleanup of soil and ground water later.

2. Scope

2.1. This procedure covers all capital and operating expenses associated with environmental control and protection.

3. Responsibilities

3.1. It is the responsibility of the environmental department to identify, to the best of their ability, environmental expenses before they occur.

3.2. It is the responsibility of upper management to make the financial resources available for necessary environmental expenditures.

3.3. The finance department is responsible for establishing and maintaining expense tracking systems.

4. Procedure

4.1. Initial Estimate and Presentation of Expenses

Expected expenses will be assessed and financial resource requirements presented to senior management. The presentation will be done to obtain budget

TITLE:	PROCEDURE NO:	Page ____ of ____

**Identification and Tracking of Financial Resources
to Cover Environmental Management**

Company Name:	Date:	*Rev.*

approval from senior management or to illustrate the financial aspects of environmental, health and safety work to the accounting department. The presentation will help to illustrate past and future environmental expenses.

42   Expense Categories To Be Established

The number of dollars in the present and planned budget will be identified for

- Environmental personnel salaries
- Environmental supplies – videos, publications, PPE, cleanup materials, labels, signs, etc.
- Proactive environmental projects – waste minimization, design for environment, etc.
- Environmental training and development
- Travel concerning environmental issues
- Contractors or vendors – consultants, lawyers, labs, temps.
- Cleanups
- Fines, penalties, litigation
- Equipment, tools

Accounting codes will be set up so that the above categories can be tracked. This will help in forecasting budgets, auditing performance and answering agency questions.

4.3.   Ongoing Tracking of Expenses

After the accounting codes are established the actual expenses will be compared to the budget and variances brought to the attention of senior management.

TITLE:	PROCEDURE NO:	Page _____ of _____

**Identification and Tracking of Financial Resources
to Cover Environmental Management**

Company Name:	Date:	*Rev.*

4.4. Establishment of Future Year Budgets

A reasonable amount of money will be identified for future budgets to protect the environment adequately.

5. Related Documentation

5.1. Accounting Records

5.2. Past Cleanup and Other Environmental Invoices

TITLE:	PROCEDURE NO:	Page _____ of _____
Dealing with Environmental Impacts		
Company Name:	Date:	*Rev.*

Company: _____

Site: _____

Dealing with Environmental Impacts

This is a "controlled" document.
Routine distribution is restricted to the approved
distribution in _____. All other persons
in possession of this document have uncontrolled copies
and should call document control for revision level status.

Approved by _____

Approved Date _____

Confidential

TITLE:	PROCEDURE NO:	Page ____ of ____
Dealing with Environmental Impacts		
Company Name:	Date:	*Rev.*

Procedure for Dealing with Environmental Impacts

1. Purpose

1.1. To identify and interpret environmental impacts.

1.2. To design cost-effective controls to minimize impacts.

2. Scope

2.1. This procedure deals with environmental impacts, effects and aspects caused by the facility.

2.2. This procedure deals with environmental impacts caused by the product or service.

2.3. This procedure deals with environmental impacts caused by contractors or vendors.

3. Responsibilities

3.1. It is the responsibility of the environmental department to identify, analyze all available information and quantify environmental impacts.

3.2. It is the responsibility of the department creating the impact to do everything it can to reduce the impact to the environment cost-effectively.

4. Procedure

4.1. Determination of Type of Effects to Analyze

There are various types of effects that should be analyzed. The highest priority includes operational impacts associated with the facility. The effects of the cur-

TITLE:	PROCEDURE NO:	Page _____ of _____
Dealing with Environmental Impacts		
Company Name:	Date:	*Rev.*

rent facility or operation will be compared with the previous operation. Effects include those associated with planned actions, start-ups, normal operations (current), incidents, shut downs and past operations. Effects of the product, service and vendors will also be considered.

A level of accuracy will be specified whenever the effects are being analyzed. For example, the accuracy of predicting the effect of a planned action is much less than the accuracy associated with a current operation. This is due to the obvious reason that a real measurement can be made of a current operation.

4.2. Perform routine audits by staff and note possible effects/exposures.

4.3. Review normal monitoring data for high exposure levels.

4.4. Review all employee complaints and suggestions.

4.5. Have special audits done by outside consultants at a set frequency and, if special concerns are identified, note possible effects/exposures.

4.6. Risk Assessment Type of Analysis of the Effects

The normal risk assessment procedure may be appropriate for an organization to use to analyze certain effects or exposures identified. The risk assessment procedure for environmental effects includes a receptor characterization step, a hazard assessment step, an exposure assessment and a risk characterization step. If this fits the particular situation, it will be used. If, on the other hand, the traditional risk assessment procedure does not make sense, a unique case-by-case analysis procedure will be used. In this case, the following common-sense procedural steps will occur:

4.7. Evaluation of the effect in terms of obvious and acute (immediate) adverse impact on employee   health and safety.

4.8. Evaluation of the effect in terms of obvious and immediate adverse impact on the closest component of the environment.

TITLE:	PROCEDURE NO:	Page _____ of _____
Dealing with Environmental Impacts		
Company Name:	Date:	*Rev.*

4.9. Evaluation of the effect in terms of compliance with regulations.

4.10. Evaluation of the effect in terms of compliance with organization standards and policies.

4.11. Evaluation of the effect in terms of whether it is controlled consistent with industry practice.

4.12. Evaluation of the effect in terms of nonobvious and chronic (long-term) impact on employees and the environment.

4.13. Prioritization of Effects

Once the above steps are completed, the effects will be prioritized so that the most significant ones can be addressed first. Most organizations do not have unlimited resources and therefore can't attack them all at the same time. The effects that are causing obvious damage to the environment or human health will be tackled first. Effects or impacts that are costing the organization considerable money will be next on the priority list.

4.14. Preparation of a Register of Effects

Effects will be organized and presented as illustrated in Figure 8-1. A separate form such as this will be used for each air, water and solid waste discharge or impact. This may result in a great number of forms, however, they would help keep impacts clearly identified and therefore easier to correct and track.

The subdivisions of the register of effects and the way the completed sheets (Figure 8-1) will be organized is as follows:

- Air emissions
- Water discharges
- Solid nonhazardous and hazardous waste
- Contamination of land
- Use of resources – land, water, fuels, energy

TITLE:	PROCEDURE NO:	Page _____ of _____
Dealing with Environmental Impacts		
Company Name:	Date:	*Rev.*

- Aesthetic effects – noise, odor, dust, vibration, and visual

- Effects on specific parts of the ecosystem

The type of operation must be considered for each of these effects. For example, effects will be identified for normal operations, start-up operations, shutdown operations, incidents, planned activities and past activities. Organizations that have several different sites may fit into several of these categories. All phases of the operation will be considered.

4.15. Control of the Effects

A procedure to select the appropriate control to address the effect will be established on a case-by-case basis. Whenever possible this procedure will specify the best available technology (BAT). The control selected will also be cost-effective, safe for the employees to operate and one that will obtain agency and public acceptance. Figure 8-2 presents a matrix or framework which will be used to evaluate several possible control options. These options will usually be different treatment technologies.

5. Related Documentation

5.1. Monitoring Data

5.2. Risk Assessments

5.3. Impact Reduction Design Documents

5.4. Regulatory Standards

5.5. Company Standard Operating Procedures

TITLE:	PROCEDURE NO:	Page _____ of _____
Identification and Achievement of Environmental Objectives and Targets		
Company Name:	Date:	*Rev.*

Company: _____

Site: _____

Identification and Achievement of Environmental Objectives and Targets

This is a "controlled" document.
Routine distribution is restricted to the approved
distribution in _____. All other persons
in possession of this document have uncontrolled copies
and should call document control for revision level status.

Approved by _____

Approved Date _____

Confidential

TITLE:	PROCEDURE NO:	Page _____ of _____
Identification and Achievement of Environmental Objectives and Targets		
Company Name:	Date:	*Rev.*

Identification and Achievement of Environmental Objectives and Targets

1. Purpose

1.1. To identify environmental objectives and targets that facilitate achievement of the environmental policy statement.

2. Scope

2.1. This procedure deals with the analysis of operational details, environmental impacts and regulatory standards and the establishment of objectives and targets to address these issues and the policy statement.

3. Responsibilities

3.1. It is the responsibility of the environmental department to analyze appropriate information and to formulate objectives and targets.

3.2. It is the responsibility of all employees to offer suggestions and to work toward achievement of the objectives and targets.

4. Procedure

4.1. Identification and Communication of Objectives and Targets

4.1.1. Using the Policy Statement

The environmental policy statement will be broken down into major subject components. At least one concrete, measurable objective and several targets for each of the subjects will be identified.

TITLE:	PROCEDURE NO:	Page ____ of ____

**Identification and Achievement
of Environmental Objectives and Targets**

Company Name:	**Date:**	*Rev.*

4.1.2. Considering the Regulations

The major site-specific legislative requirements, such as obtaining a permit to initiate an operation, will be identified. These will also be made objectives and a target in terms of timing will be specified.

4.1.3. Considering the Impacts

An objective to reduce major impacts of the operation associated with the larger hazardous waste streams will also be identified. An annual percent reduction will be a target.

4.1.4. Considering the Organization's Operational Goals

At least one of the organization's top goals that must be achieved to stay in operation, such as a certain profit margin, will be identified as an objective. A determination of how the environmental department can help support this goal, such as a target of 10 percent reduction in environmental expenses in the next year, will be stated.

4.1.5. Considering the Input of Interested Parties

An attempt will be made to solicit and record views of interested parties. The unsolicited comments expressed by customers, agencies, environmental groups or other individuals will also be considered. A determination will be made whether the suggestions will truly help the environment. If so the reasonable views will be made into objectives and targets.

4.1.6. Communication of the Objectives and Targets

The objectives and targets will be communicated to appropriate individuals in the organization.

4.1.7. Modification of the Objectives and Targets

The objectives and targets will be modified when changes occur in the organization or to the product.

TITLE:	PROCEDURE NO:	Page _____ of _____

**Identification and Achievement
of Environmental Objectives and Targets**

Company Name:	**Date:**	*Rev.*

4.2. Procedure for Achievement of the Objectives and Targets

4.2.1. Prepare an Action Plan

In order to help achieve the objectives and targets a separate action plan for each will be prepared. Figure 10-1 is an example of an action plan. Note that all the major steps or actions have been identified along with a person to perform the work. This designation of responsibility for achieving objectives and targets is essential. Also of critical importance are the completion or due dates for each step. When problems are encountered during execution of some of the steps, and there will be problems, the action plan must be adjusted if the target is going to be achieved.

4.2.2. Use PERT Charts or Other Techniques If Necessary

For complicated objectives an action plan alone may not be enough. In these cases it is best to also utilize tools, such as project evaluation and review techniques (PERT). PERT systems/charts show all the interrelated tasks, along with early and late finishes, critical paths and other important information necessary to achieve the targets. All of the horizontal lines in a PERT chart are individual tasks necessary to meet the target. The vertical lines illustrate interrelations among the tasks. The heavy line is the critical path, which must be completed on time or the entire target date will be missed. After input of certain data, some software systems will not only print out the PERT chart itself but also a status sheet for each subtask, along with early start, early finish, late start, late finish and other information.

4.2.3. Dealing with Changes

Objectives and targets should be adjusted when environmental and business conditions warrant. This will probably occur much more than a person would like, but in many cases changes are unavoidable. There needs to be a balance between too many changes to targets and objectives and no changes. Either end of the spectrum can spell the end to meaningful and cost-effective environmental management.

TITLE:	PROCEDURE NO:	Page _____ of _____

**Identification and Achievement
of Environmental Objectives and Targets**

Company Name:	Date:	*Rev.*

In order to preserve the integrity of the overall environmental management program, the changes should be made from the microlevel and progress, if necessary, to the macrolevel. The first change should be to subtasks in the action plan for one objective. If this is not enough, then an adjustment should be made to several subtasks or an entire target. If all else fails, an entire objective may have to be upgraded or even replaced. When macro level changes are made to whole objectives or targets, it is very important to consider whether the change will impact other objectives and even the overall policy statement. This can happen in some situations due to the complex interrelationship of environmental issues.

4.2.4. Designation of Responsibility for Meeting Targets

Overall, it is important to have as much employee involvement in the decision making as possible. Flexibility in the achievement of targets and ability to make adjustments to the process involved are perfect examples of the responsibility the employees need. This requires that upper management trust and support their employees to a much greater extent than is normally found in real life.

If responsibilities are not assigned, the targets will probably not be achieved. Therefore it is very important to designate key individuals at each organizational level who play a part in completion of the objective or target. For example, the president of the organization would be responsible for signing the environmental policy statement and supporting the concepts it promotes. The director or officer in charge of the environmental management department would be responsible for obtaining the human and financial resources necessary for complying with the policy statement and developing the objectives with input from as many employees as physically possible. Environmental manager(s), engineer(s) or specialist(s) should be assigned responsibility for specific target(s).

Completion of most objectives and targets will also require assigning responsibility for completion of certain action steps to individuals outside of the environmental department. For example, individuals in the following departments should be assigned responsibility for helping to meet environmental objectives and targets:

TITLE:	PROCEDURE NO:	Page _____ of _____
Identification and Achievement of Environmental Objectives and Targets		
Company Name:	Date:	*Rev.*

- Environmental Department – lead or coordination of entire objective
- Engineering Department – implementation of many waste minimization projects
- Procurement/Purchasing Department – purchase of hazardous material services or products
- Legal Department – review of environmental laws and contracts
- Human Resources Department – assistance with environmental training
- Public Affairs Department – assist with response to media and customer questions

4.3. Objective and Target Tracking System

Many targets may have minimum quantity or quality outcomes specified. If this is the case these criteria or standards must be identified and tracked so that it will be obvious when the objectives and targets have been met. Figure 10-2 illustrates a possible tracking system for objectives and targets.

5. Related Documentation

5.1. Environmental Policy Statement

5.2. Impact Assessment

5.3. Regulatory Standards

TITLE:	PROCEDURE NO:	Page _____ of _____
Record and Document Control Procedure		
Company Name:	**Date:**	*Rev.*

Company: _____

Site: _____

Record and Document Control Procedure

This is a "controlled" document.
Routine distribution is restricted to the approved
distribution in _____. All other persons
in possession of this document have uncontrolled copies
and should call document control for revision level status.

Approved by_____

Approved Date _____

Confidential

TITLE:	PROCEDURE NO:	Page _____ of _____
Record and Document Control Procedure		
Company Name:	Date:	*Rev.*

Record and Document Control Procedure

1. Purpose

1.1. To identify, retain and then systematically destroy environmental records at the appropriate time.

2. Scope

2.1. This procedure covers all environmental related documentation.

2.2. This procedure covers paper and electronic forms of documentation.

3. Responsibilities

3.1. It is the responsibility of the environmental department to identify all types of environmental records, documentation and retention periods required by environmental regulations.

3.2. It is the responsibility of the legal department to determine the retention periods of all types of documents with other legal specified retention times.

3.3. It is the responsibility of all individuals involved with these documents to adhere to the specifications.

4. Procedure

4.1. Identification of Appropriate Documents for Retention

A first step in document control is a thorough identification of what needs to be retained for regulatory and business reasons. Table 16-1 lists many of the environmental documents which should be retained and an estimate of the amount of time to keep each document. For example, this list shows that the

TITLE:	PROCEDURE NO:	Page _____ of _____
Record and Document Control Procedure		
Company Name:	**Date:**	*Rev.*

manifest is to be kept for three years. The Resource Conservation and Recovery Act (RCRA) requires this and an organization can be fined up to $25,000 per day if it is not done. Many of the other dates shown in Table 16-1 are not required by regulation but just make good business sense. If a fine is assessed or litigation develops, it may be important to have copies of these documents.

4.2. Collection of Documents

One obvious way to collect documents is to instruct individuals who may be involved with the document to submit a copy to you. Sometimes this is not too reliable, and it is, therefore, better to either pick one up yourself or be involved with the original document and make your own copy. Close supervision to assure compliance is needed.

4.3. Indexing of Documents

Table 16-1 is arranged to illustrate one way that environmental documents can be indexed. There are, however, an infinite number of other ways to organize the documents. Whichever way is selected, it should be computer-based to permit easier search by subject.

4.4. Filing and Storing of Documents

Environmental files need to be well organized, kept up to date and locked. Part of the reason for this is the high number of environmental lawsuits. It is also a good idea for the documents to be retained in either a fire-rated room, fire-rated file cabinets, or a backup copy kept at another location. If the volume of files exceeds the space available, it may be necessary to have a set of active file cabinets on-site and inactive files off-site.

4.5. Removal of Obsolete Files

Once the time listed on Table 16-1 is reached, it is important to promptly destroy the outdated files for two reasons. First, old paperwork can become a nightmare during litigation, since it would probably have to be produced if it

TITLE:	PROCEDURE NO:	Page ____ of ____
Record and Document Control Procedure		
Company Name:	Date:	*Rev.*

still exists. Also, most organizations have limited space and purging old files will free up space for more important documents. For the removal of obsolete files to be accomplished promptly and efficiently, the computer-based indexing system mentioned above will be of value.

4.6. Maintenance of Document Control Specifications

The same document control specifications required for ISO 9000 certification apply here as well (refer to Table 16-2). For example, each page will have numbers (such as page 1 of 3). All documents will have issue date, retention time and revision numbers. If the document is related to another it must be cross-referenced. Again, since the specifications are the same, they will not be repeated here.

4.7. Preparation and Distribution of Entire Controlled Documents

The procedure for preparing and distributing documents will be followed. As an illustration, the following general steps apply to a standard operating procedure binder for environmental, health and safety. The most important goal of this entire process is controlled preparation and distribution of the environmental procedures to the proper individuals.

4.7.1 The entire binder will be prepared by the corporate environmental, health and safety (EHS) department with input from others as appropriate.

4.7.2. At a minimum, one numbered and recorded copy will be given to the lead site individual, the senior-most environmental, health and safety individual, the safety committee chair and the training coordinator

4.7.3. All requests for a copy should be directed to the corporate EHS department.

4.7.4. After the individual receives a copy of the binder, they must send back the receipt sheet verifying that they received the binder, have completed a review, understand its contents and will comply.

TITLE:	PROCEDURE NO:	Page _____ of _____
Record and Document Control Procedure		
Company Name:	**Date:**	*Rev.*

4.8. Preparation and Distribution of Controlled Individual Procedures

> **4.8.1.** After an individual procedure has been revised or prepared for the first time, it must be forwarded to the corporate EHS department for checking and inclusion in the controlled binder.

> **4.8.2.** Corporate EHS will forward approved individual procedures to each binder holder, along with an updated index.

4.9. Preparation and Distribution of Uncontrolled Individual Procedures

It may not be cost-effective or necessary for all employees to have an entire controlled binder of procedures. For example, certain administrative employees may need only to understand a couple procedures. In these cases, the following procedure will be followed:

> **4.9.1.** The site environmental, health and safety individual, supervisor or any other site individual who has a controlled binder may make uncontrolled copies of individual procedures, as needed.

> **4.9.2.** Before the copy is given out it must be stamped "Uncontrolled." The recipient should be told that they are not on the distribution list for updates since only controlled binder holders receive revisions and updates.

> **4.9.3.** The controlled binder holder should keep a list of what uncontrolled procedure copies they have distributed, and to whom, so that revisions and updates can be forwarded if appropriate. In real life this probably will not happen unless done by the central coordinating body.

4.10. General Distribution Guidelines

Before documents are distributed a determination of whom should receive them is made first. Figure 16-1 shows a distribution matrix for one type of document – environmental, health and safety standard operating procedures (SOPs). The figure illustrates that certain populations of employees should get only certain SOPs and some employees should get all SOPs.

TITLE:	PROCEDURE NO:	Page ____ of ____
Record and Document Control Procedure		
Company Name:	Date:	*Rev.*

As with document control, the distribution specifications are the same as they were for ISO 9000. For example, controlled copies sent out are to be numbered, whereas uncontrolled copies are not numbered. Only the controlled copy holders receive updates. When certain important documents are received, a return receipt should be sent back to the distributor. Figure 16-2 shows one possible form for ensuring that documents have been received.

4.11. Special Protection of Certain Documents

Certain documents and verbal communications need special protection, not because they deal with illegal issues, but because it is inappropriate material for the court room. For example, some of the first paperwork or statements made concerning an environmental problem will probably be full of guesses, erroneous theories, errors, assumptions, value statements and other incorrect and misleading information. If this material is not protected under Attorney-Client Privilege, it can legally be "discovered" or brought to court and result in days of wasted time going down an incorrect path. This is not to the best interest of anyone involved.

The best way to implement Attorney-Client Privilege is by getting the attorney involved early. If the attorney requests files, samples, and other information, the material will be prepared at the request of counsel and protected under Attorney-Client Privilege. All documents generated in this way should be clearly marked "Attorney-Client Privilege." The following list suggests when to possibly consider implementing Attorney-Client Privilege:

4.11.1. The public becomes involved in a sensitive internal issue.

4.11.2. Negligence or criminal conduct on the part of an employee is suspected.

4.11.3. A regulatory violation has probably occurred.

4.11.4. Significant environmental damage has probably occurred. Involving the attorney should not, however, delay action on your part to help contain the situation or minimize the impact. For example, if a spill has occurred that involves public property, sewers, soil, storm drains or air quality, it has to be

TITLE:	PROCEDURE NO:	Page ____ of ____
Record and Document Control Procedure		
Company Name:	**Date:**	*Rev.*

contained and reported. Once the immediate crisis is under control, call the attorney.

4.11.5. Significant health and safety impact has occurred. Proceed the same as in 4.11.4.

4.11.6. A lawsuit has been threatened

4.11.7. Site investigations where air, water or soil is sampled and contamination is found to be above regulatory limits

4.11.8. A notice of violation or other notice of noncompliance is received

4.11.9. An attorney contacts the organization on behalf of someone else

4.11.10. The media is involved in a negative event

4.11.11. A special interest group is involved in a negative event

4.11.12. The police, FBI or district attorney are involved.

5. Related Documentation

5.1. All Records and Documents Mentioned in this Procedure

TITLE:	PROCEDURE NO:	Page _____ of _____
Handling Regulatory Requirements		
Company Name:	Date:	*Rev.*

Company Name: _____

Site: _____

Handling Regulatory Requirements

This is a "controlled" document.

Routine distribution is restricted to the approved

distribution in _____. All other persons

in possession of this document have uncontrolled copies

and should call document control for revision level status.

Approved by _____

Approved Date _____

TITLE:	PROCEDURE NO:	Page _____ of _____
Handling Regulatory Requirements		
Company Name:	Date:	*Rev.*

Procedure for Handling Regulatory Requirements

1. Purpose

1.1. To identify and comply with applicable regulatory requirements.

2. Scope

2.1. This procedure covers the identification, analysis and actions needed to comply with applicable regulatory requirements.

2.2. This procedure covers regulations at the international, national/federal, state, regional and local levels of government.

3. Responsibilities

3.1. It is the responsibility of the environmental department to identify and analyze environmental regulations.

3.2. The legal department will assist with regulatory interpretations.

3.3. It is the responsibility of all employees to comply with the regulations.

4. Procedure

4.1. Determination of the Type of Regulatory Requirements to Be Identified

Any regulation that applies to the operation will be identified. This may include international, federal, state, regional and local regulations. At each level of government there will be several different regulations.

Types of regulatory requirements which will be identified are of all government levels and in all segments of the environment. This includes legislation that covers the protection of air, water, land and natural resources.

TITLE:	PROCEDURE NO:	Page _____ of _____
Handling Regulatory Requirements		
Company Name:	Date:	*Rev.*

4.2. Impacts

A list of major areas of the environment impacted by the operation, that is, air, water, soil, wildlife, land, and so forth, will be made. The list will also include major types of waste that the operation generates, such as hazardous waste, paper, aluminum, glass, and so forth.

4.3. Agency Lists

A check of the numerous agency lists that are available will be made. If the operation impacts a component of the environment or generates waste, the corresponding agency should be consulted since they probably have regulations that apply. If a commercially prepared agency list is not available, a check of the phone book under federal, state, county and city governments will be made. Within each of those major sections, there will probably be specific agencies arranged by subject.

4.4. Regulatory Services

A regulatory service, library or a contact to the agency will be made to get a copy of their regulations. They will be asked if they know of other regulations or agencies that might pertain. All levels of government will be checked to see if they have regulations that apply, including international, federal, state, regional and local.

4.5. Index and Preamble

The index and preamble of the regulations will be checked and then the relevant section. The background sections will be bypassed if the main body of the regulation does not apply.

4.6. Reading

Speed reading the applicable section of the regulation will be done next and if it even remotely correlates to the operation, more analysis of the regulation will occur.

TITLE:	PROCEDURE NO:	Page _____ of _____
Handling Regulatory Requirements		
Company Name:	Date:	*Rev.*

4.7. Regulation Summary

A one-page regulatory summary will be prepared (see Figure 9-1). This summary sheet will show whether the regulation is international, federal, state, regional or local. Since most regulations are lengthy and complicated, the one-page summary sheet is valuable as a reminder and as an executive summary for upper management. These sheets will also be completed for proposed regulations as well, especially if they will have a big impact on the organization and are likely to pass.

4.8. Filing of the Summary Sheets

There are various ways to organize the regulation summary sheets or the actual regulations themselves. This would involve grouping all international regulations together, followed by federal, state and local. A notation on Figure 9-1 should indicate whether the paper file, binder or computer file contains the regulation and also whether it is physically located at corporate headquarters, the operating site or consultant location.

5. Related Documentation

5.1. Regulations

5.2. Laws and Acts

5.3. Regulatory Summaries

5.4. Monitoring Data to Illustrate Compliance

5.5. Permits and Permit Applications

5.6. Plans Required by Regulations

TITLE:	PROCEDURE NO:	Page _____ of _____
Environmental Training Procedure		
Company Name:	Date:	*Rev.*

Company Name: _____

Site: _____

Environmental Training Procedure

This is a "controlled" document.

Routine distribution is restricted to the approved

distribution in _____. All other persons

in possession of this document have uncontrolled copies

and should call document control for revision level status.

Approved by _____

Approved Date _____

Confidential

TITLE:	PROCEDURE NO:	Page _____ of _____
Environmental Training Procedure		
Company Name:	Date:	*Rev.*

Environmental Training Procedure

1. Purpose

1.1. To identify, provide and track environmental training that will help employees to minimize impacts.

2. Scope

2.1. This procedure covers the identification of training requirements for all employees in the organization.

2.2. This procedure covers the types, methods and frequency of environmental training.

3. Responsibilities

3.1. Training needs should be identified by the environmental department.

3.2. Training should be done by the training department, environmental department or outside vendor.

3.3. Overall responsibility and tracking of environmental training is the responsibility of the environmental department.

4. Procedure

4.1. Identification of Environmental Training Needs

This procedure will identify training needs of the individual. The scope and depth of training needs to be appropriate. Identification of nonhazardous, chemical and hazardous wastes that employees might encounter is the first step.

TITLE:	PROCEDURE NO:	Page _____ of _____
Environmental Training Procedure		
Company Name:	**Date:**	*Rev.*

4.2. Identification of Laws and Regulations Requiring Training

Identification of laws that require training for the chemicals and hazardous wastes identified (Occupational Safety and Health Act, Resource Conservation and Recovery Act, Hazardous Material Transportation Act, etc.) is the next step in the procedure.

4.3. Matching of Employee's Abilities with Required Training

Next there should be a match of employees (by name or job type) with the required training. An example of how this identification might look is presented in Table 7-2, Training Matrix. This table shows different training classes across the top and nine categories of employee along the side. A • is placed in the rows to indicate that the training is required for that particular employee type. This table also shows the recommended frequency of the training and length.

4.4. Providing the Training

Adequate resources are to be available to provide the training identified. Some training will be provided by certified trainers, such as first aid and CPR. As with corrective actions identified in an audit, training needs can't just be identified and not filled. The training will be provided within a certain period of time from hiring and some annual refresher provided as well. The environmental training will be provided by a variety of resources, depending on size and type of employee body. When there is a large number of employees who need the specific type of training, it is more cost-effective to have an internal training department provide much of the training. When the population of employees needing the training is smaller it will be done by the environmental manager. The corporate environmental department will provide certain types of training depending on their own workload. Consultants will provide on-site or "suitcase" training or off-site training. HAZWOPER training is an example of training usually provided by consultants. Whatever the mix of training resources used it will be noted on a matrix with footnotes such as:

TITLE:	PROCEDURE NO:	Page _____ of _____
Environmental Training Procedure		
Company Name:	Date:	*Rev.*

1 = Environmental Department Provided

2 = Training Department Provided

3 = Consultant Provided

4.5. Tracking the Training Hours

To show compliance with the laws mentioned above the training hours will be collected and tracked. It is not enough to just keep an attendee list for the file. In addition to name, the employee's social security number, job title, job description and training requirements will be kept in the file. Total training hours by category will be calculated and reported to management. Hours per employee will be determined to ensure that goals are met. If possible, a computer database that matches employee name and job code with required training will be used. A printout would then show whether the particular employee is adequately trained and current.

5. Related Documentation

5.1. Training Records

5.2. Instructor Certification Records

5.3. Instructor Packages

TITLE:	PROCEDURE NO:	Page _____ of _____
Environmental Procedures Concerning Procurement and Vendor Control		
Company Name:	Date:	*Rev.*

Company Name: _____

Site: _____

Environmental Procedures Concerning Procurement and Vendor Control

This is a "controlled" document.
Routine distribution is restricted to the approved
distribution in _____. All other persons
in possession of this document have uncontrolled copies
and should call document control for revision level status.

Approved by _____

Approved Date _____

Confidential

TITLE:	PROCEDURE NO:	Page _____ of _____
Environmental Procedures Concerning Procurement and Vendor Control		
Company Name:	**Date:**	*Rev.*

Environmental Procedures Concerning Procurement and Vendor Control

1. Purpose

1.1. To insure that suppliers/vendors/contractors for the organization are doing their part to protect the environment.

2. Scope

2.1. This procedure covers vendors/contractors/suppliers that provide materials, parts, chemicals and fuels.

2.2. This procedure covers vendors/contractors/suppliers that provide environmental and non-environmentally related services.

3. Responsibility

3.1. Environmental considerations, restraints or issues are to be identified and analyzed by the environmental department.

3.2. The procurement/purchasing department is responsible for obtaining contractor information, the minimum number of quotes from qualified vendors and for negotiating a good price and contract.

3.3. The user of the part, chemical or service is responsible for making sure that whatever is delivered is in accordance with the agreement.

TITLE:	PROCEDURE NO:	Page _____ of _____
Environmental Procedures Concerning Procurement and Vendor Control		
Company Name:	Date:	*Rev.*

4. Procedure

4.1. Prequalification Step

A collection of the credentials of as many vendors as physically possible will be made. This will include their experience, financial stability, insurance, permits, resumes of key employees, quality programs and other information. The material is then screened, questioned, and supplemented and a determination is made whether the vendor is qualified to perform the environmental service or supply the product containing a hazardous material. Screening criteria will be established so that all vendors are treated equally. Some suggested screening criteria are shown in Table 11-2. If these criteria are used for all vendors considered, then a fair comparison can be made. Probably the hardest of these criteria to quantify is the first one shown: experience. A shorter period of relevant experience is usually better than more years of general experience.

This prequalification step applies to all environmentally related vendors. For example, it should be done for environmental consulting service vendors and chemical or hazardous material sale vendors. Prequalification should also be done for hazardous waste treatment, storage and disposal facilities (TSDFs). In the case of the latter, it is essential to visit the TSDF site and do an annual audit of the facility (see Appendix C). A check of the agency files in reference to the TSDF is also important.   Once the qualification step has been completed for a particular vendor, they should be added to the "approved vendor" list if they meet the qualification criteria. See Figure 11-1 for a sample format. All sites should consider only vendors from the list. It is important to keep reassessing the vendors and adjusting the list frequently, as their status can change quickly.

4.2. Request for Proposal or Quote

Only the vendors that meet the screening criteria mentioned above and are on the approved list will be invited to submit a proposal or quote. If the service or product is clearly understood by both the buyer and the seller, a quote is probably sufficient. If, however, the environmental service is complicated or not well

TITLE:	PROCEDURE NO:	Page _____ of _____

**Environmental Procedures Concerning
Procurement and Vendor Control**

Company Name:	**Date:**	*Rev.*

spelled out, it is best to ask all qualified vendors for a proposal. The proposal should detail the steps the vendor will follow to complete the job.

The request for a proposal or quote should be made in writing and strict documentation controls followed. For example, all vendors should get identical requests which are mailed at the same time. If the vendors have questions, they should all receive the same answers at the same time in writing.

Usually vendors are given the opportunity to see the site or situation (walkthrough). All vendors should visit the site at the same time so that everyone has the same opportunity and hears all questions and answers. The quote or proposal must then be submitted in writing and received in the specified time or it should be disqualified.

4.3. Review of Vendor Submittal

Clear and measurable criteria for selecting the successful vendor must be in writing. There is a strong chance that this step might be questioned by an unsuccessful vendor. Some suggested criteria were shown in Figure 11-1 for selecting a hazardous waste service vendor. The criteria should concentrate on the proposal, not the vendor's general qualifications. It is assumed at this point that all vendors invited to send a quote or proposal are already prequalified.

Obviously, one major selection criterion is cost. Far too often only the cost is considered and the job awarded to the low quoter. In some cases this can be a mistake as some vendors come in with a low quote and make it up later with add-ons.

4.4. Selection of Vendor and Execution of Agreement

All vendors will be informed of the outcome and the successful vendor needs to sign the agreement covering the service or product. The draft agreement should have been part of the request for proposal or quote sent earlier so that last minute agreement surprises or negotiations will not occur. Written agreements are an absolute requirement and can be either a one-job type agreement or a master service type of agreement. Whichever type it is, liabilities must be clearly

TITLE:	PROCEDURE NO:	Page _____ of _____
Environmental Procedures Concerning		
Procurement and Vendor Control		
Company Name:	**Date:**	*Rev.*

spelled out and fair to both the buyer and the vendor. Also certain of the buyer's environmental policies must be included in the agreement so that the vendor knows to follow them when they start work or shipping product.

4.5. Initiation of the Work

Once all the agreement terms and conditions are worked out the environmental work can be started by the vendor or the product containing a hazardous material shipped. Right from the start the vendor must adhere to the buyer's agreement terms, especially compliance with environmental policy.

4.6. Maintaining a Good Ongoing Vendor Relationship

A good ongoing relationship is important with environmental vendors, as it is with all other types of vendors. You don't want an environmental vendor cutting corners because they quoted too low or because they did not understand the original scope of the job.   Also, if a big liability problem develops, no matter what the agreement states, both the buyer and the vendor are going to suffer. A team effort or partnership with the vendor during the project will reduce the number of problems for both parties.

5. Related Documentation

5.1. Vendor Agreements

5.2. Request For Quotation rr Proposal Documents

5.3. Material Safety Data Sheets

5.4. Product or Job Specifications

TITLE:	PROCEDURE NO:	Page _____ of _____
Management of Nonhazardous Materials and Wastes		
Company Name:	Date:	*Rev.*

Company Name: _____

Site: _____

Management of Nonhazardous Materials and Wastes

This is a "controlled" document.
Routine distribution is restricted to the approved
distribution in _____. All other persons
in possession of this document have uncontrolled copies
and should call document control for revision level status.

Approved by _____

Approved Date _____

Confidential

TITLE:	PROCEDURE NO:	Page _____ of _____
Management of Nonhazardous Materials and Wastes		
Company Name:	Date:	*Rev.*

Procedure for Management of Nonhazardous Materials and Wastes

1. Purpose

1.1. To preserve valuable landfill space and natural resources.

1.2. To promote waste minimization.

2. Scope

2.1. In this procedure, paper, aluminum, glass, cardboard, and plastic will be covered in terms of waste minimization.

2.2. For all these materials the four Rs apply, that is, reduce, reuse, recycle and repurchase.

3. Responsibility

3.1. It is the responsibility of the environmental department to set up a waste minimization program and oversee its progress, collect data, communicate to employees and upgrade the program.

3.2. It is the responsibility of every employee to do their part to reduce, reuse and recycle waste materials wherever they can.

3.3. It is the responsibility of the facility department to make arrangement with vendors for the recycling of waste material.

TITLE:	PROCEDURE NO:	Page _____ of _____
Management of Nonhazardous Materials and Wastes		
Company Name:	Date:	*Rev.*

4. Procedure

4.1. Reduction

If everyone tries to reduce the amount of materials they use, there will be less volume to handle. For example, if double sided photocopying is done, less paper will have to be purchased. Use of E mail or other electronic systems, computer files and restricted distributions will also reduce the amount of paper used. The same principles apply to the other materials.

4.2. Reuse

By reusing paper, cartons and packing and other materials, the volume of waste to be handled later will be reduced. If the material can't be reused for the original purpose because of quality reasons, perhaps it can be used in some other way. For example, it may not be possible to reuse a carton to ship new product to a customer, however, the carton could be reused for shipping defective parts back to a supplier. By reusing materials to the maximum extent possible, the organization is saving resources, cost and valuable landfill space.

4.3. Recycling

4.3.1. Materials Premarked for Recycling

Some materials, such as plastic, have recycle symbols already printed on the new product. It may even be possible for the user organization to mark some recycle symbols on supplies and other products, in some cases, if they do not already contain a designation. These symbols help the user know what is recyclable.

4.3.2. Inventory

An accounting of types and volumes of materials to be recycled should be made. This may involve collecting figures from all of the organization's sites in order to achieve economical volumes. The figures can be estimates but should be conservative. In some parts of the world, such as Germany, law requires that

TITLE:	PROCEDURE NO:	Page _____ of _____
Management of Nonhazardous Materials and Wastes		
Company Name:	**Date:**	*Rev.*

the manufacturer take back packing materials for recycling. This volume must also be added into the inventory.

4.3.3. Contract with Recycler

An agreement should next be set up with one or more recyclers. Before signing the agreement, it would be a good idea to talk to several recyclers so that the best deal can be obtained. Some governments, such as Germany, may require the use of a specified recycler (Green Dot Program). Details such as who provides the collection containers, dates of collection and charges or credits are reflected in the contract. Another detail that is important to work out is whether all materials will be put into separate recycling bins or one container and sorted at the recycler facility. If space permits, it is usually better to have separate containers at the generator's location. This would mean separate containers for paper, cardboard, plastic, glass and aluminum.

4.3.4. Placement of Recycle Containers

Unfortunately, placement of the recycle container in relation to the generators of the waste materials plays a big part. If it is too far away or confusing to use, many people will not take the time to recycle. The containers should be placed as close to the generators or source of the waste as possible. They should be clearly marked, especially if the materials to be recycled are to be placed into different containers.

If there is a "Take Back" type law in place, such as in Germany, a separate container may have to be set up for packing material and used product which has been returned by the customer. It is especially important that this material be accurately accounted for and proof of recycling received.

4.3.5. Pickup and Recycling by the Vendor

When the materials are picked up, it is important to get a couple records. In addition to a weight receipt, some sort of certificate or notice that the material will be recycled is important to have.

TITLE:	PROCEDURE NO:	Page _____ of _____
Management of Nonhazardous Materials and Wastes		
Company Name:	**Date:**	*Rev.*

4.4. Repurchase

To complete the cycle or loop it is important that the organization purchase recycled products, even if they cost a little more. The purchasing department may need to set up volume discounts to help reduce this cost. If more organizations would purchase post-consumer products or recycled products it would improve their marketability and price. This is what is really needed to make recycling work.

4.5. General Procedures to Ensure That All the Programs Work

There are some common elements that will help all the programs be successful whether it be reduction, reuse, recycle or repurchase. Actually these suggestions would maximize the success of any environmental program. This includes the following:

4.5.1. Employee Awareness Program and Reminders

Most people need to be "energized" about reuse and recycling and continually reminded. It is just not at the top of the priority list for many busy individuals. Reminders can be in the form of fliers, posters, verbal announcements and other means. Waste minimization contests can be valuable as a reminder to recycle.

4.5.2. Tracking and Auditing

As with most other environmental programs, it is necessary to track and audit for compliance or progress. In this case the auditing could be done by the site's waste minimization committee. If the site is very small and does not have such as committee, then the site's environmental representative should do the audit. Corporate should also audit on occasion.

4.5.3. Targets

Each site should set some targets up along with some corporatewide goals. As was discussed previously, the targets would be actual numbers, such as pounds reduced.

TITLE:	PROCEDURE NO:	Page _____ of _____
Management of Nonhazardous Materials and Wastes		
Company Name:	**Date:**	*Rev.*

5. Related Documentation

5.1. Recycling Agreements

5.2. Certificates of Recycling

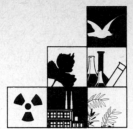

Commonly Used Abbreviations and Acronyms

ANSI	American National Standards Institute
BAT	Best Available Technology
BS	British Standard
BSI	British Standard Institute
CAS	Chemical Abstract System
CEI	Continuous Environmental Improvement
CEN	Comite European de Normalisation
CIA	Chemical Industries Association
DOT	Department of Transportation
DRT	Disaster Recovery Team
EHS	Environmental, Health and Safety
EMAS	Eco-Management and Audit Scheme
EMS	Enironmental Management System
EPA	Environmental Protection Agency
ER	Emergency Response
ERT	Emergency Response Team
EU	European Union
HAZWOPER	Hazardous Waste Operations and Emergency Response
HMW	Hazardous Material and Hazardous Waste
HR	Human Resources Department
IEC	International Electrotechnical Commission

ISO	International Organization for Standardization
MSDS	Material Safety Data Sheet
ODS	Ozone Depleting Substances
OSHA	Occupational Safety and Health Administration
PERT	Project Evaluation and Review Technique
PPB	Polybrominated Biphenyls
PPE	Personal Protective Equipment
ppm	Parts Per Million
QC	Quality Control
RMPP	Risk Management Prevention Plan
RQ	Reportable Quantity
SAGE	Strategic Advisory Group for the Environment
SARA	Superfund Authorization and Reauthorization Act
SC	Subcommittee For ISO 14000
SCBA	Self-Contained Breathing Apparatus
SHEMS	Safety, Health and Environmental Management Systems
SOP	Standard Operating Procedure
SUB TAG	Sub Technical Advisory Group for ISO 14000
TAG	Technical Advisory Group for ISO 14000
TC	Technical Committee for ISO 14000
TQEM	Total Quality Environmental Management
TSDF	Treatment, Storage and Disposal Facility
UNCED	United Nations Conference on the Environment and Development
WG	Working Group for ISO 14000

References

British Standards Institute. 1994. *Specification for Environmental Management Systems/BS 7750*. Standards Board.

Burhenn, David W. 1995. *Briefing on ISO International Environmental Standards*. Los Angeles: Sidley and Austin.

Cutter Information Corporation. 1994. Will There be Dueling European and International Management Standards? *Business and the Environment*.

Environment Today. 1993. ISO Work Groups Begin Tackling Enviro-standards. *Environment Today*. December 1993.

Morrison and Foerster. 1994. *Land Use and Environmental Law Briefing—International Standards for Environmental Issues—Is Europe Moving Ahead of the U.S.?* San Francisco: Morrison and Foerster Law Firm.

Morrow, M. *Chemical Week*. April 6, 1994.

National Center for Manufacturing Sciences (NCMS). 1994. *Focus—Exceeding Partner Expectations*. April 1994.

Rothery, Brian. 1993. *BS 7750—Implementing the Environment Management Standard and the EC Eco-Management Scheme*. Brookfield, VT: Gower Press.

SRI International. 1994. *Environmental Strategies—Design For Environment*. A Newsletter from SRI International. Spring 1994.

U.S. Sub Technical Advisory Group 1 to Technical Committee 207. 1995. Unpublished and uncopyrighted committee paperwork.

Wortham, Sarah. 1993. International: Get Ready for Worldwide Standards. *Safety and Health*. December 1993.

Index

LICENSE AGREEMENT AND LIMITED WARRANTY

READ THE FOLLOWING TERMS AND CONDITIONS CAREFULLY BEFORE OPENING THIS DISK PACKAGE. THIS LEGAL DOCUMENT IS AN AGREEMENT BETWEEN YOU AND PRENTICE-HALL, INC. (THE "COMPANY"). BY OPENING THIS SEALED DISK PACKAGE, YOU ARE AGREEING TO BE BOUND BY THESE TERMS AND CONDITIONS. IF YOU DO NOT AGREE WITH THESE TERMS AND CONDITIONS, DO NOT OPEN THE DISK PACKAGE. PROMPTLY RETURN THE UNOPENED DISK PACKAGE AND ALL ACCOMPANYING ITEMS TO THE PLACE YOU OBTAINED THEM FOR A FULL REFUND OF ANY SUMS YOU HAVE PAID.

1.   **GRANT OF LICENSE:** In consideration of your payment of the license fee, which is part of the price you paid for this product, and your agreement to abide by the terms and conditions of this Agreement, the Company grants to you a nonexclusive right to use and display the copy of the enclosed software program (hereinafter the "SOFTWARE") on a single computer (i.e., with a single CPU) at a single location so long as you comply with the terms of this Agreement. The Company reserves all rights not expressly granted to you under this Agreement.

2.   **OWNERSHIP OF SOFTWARE:** You own only the magnetic or physical media (the enclosed disks) on which the SOFTWARE is recorded or fixed, but the Company retains all the rights, title, and ownership to the SOFTWARE recorded on the original disk copy(ies) and all subsequent copies of the SOFTWARE, regardless of the form or media on which the original or other copies may exist. This license is not a sale of the original SOFTWARE or any copy to you.

3.   **COPY RESTRICTIONS:** This SOFTWARE and the accompanying printed materials and user manual (the "Documentation") are the subject of copyright. You may <u>not</u> copy the Documentation or the SOFTWARE, except that you may make a single copy of the SOFTWARE for backup or archival purposes only. You may be held legally responsible for any copying or copyright infringement which is caused or encouraged by your failure to abide by the terms of this restriction.

4.   **USE RESTRICTIONS:** You may <u>not</u> network the SOFTWARE or otherwise use it on more than one computer or computer terminal at the same time. You may physically transfer the SOFTWARE from one computer to another provided that the SOFTWARE is used on only one computer at a time. You may <u>not</u> distribute copies of the SOFTWARE or Documentation to others. You may <u>not</u> reverse engineer, disassemble, decompile, modify, adapt, translate, or create derivative works based on the SOFTWARE or the Documentation without the prior written consent of the Company.

5.   **TRANSFER RESTRICTIONS:** The enclosed SOFTWARE is licensed only to you and may <u>not</u> be transferred to any one else without the prior written consent of the Company. Any unauthorized transfer of the SOFTWARE shall result in the immediate termination of this Agreement.

6.   **TERMINATION:** This license is effective until terminated. This license will terminate automatically without notice from the Company and become null and void if you fail to comply with any provisions or limitations of this license. Upon termination, you shall destroy the Documentation and all copies of the SOFTWARE. All provisions of this Agreement as to warranties, limitation of liability, remedies or damages, and our ownership rights shall survive termination.

7.   **MISCELLANEOUS:** This Agreement shall be construed in accordance with the laws of the United States of America and the State of New York and shall benefit the Company, its affiliates, and assignees.

8.   **LIMITED WARRANTY AND DISCLAIMER OF WARRANTY:** The Company warrants that the SOFTWARE, when properly used in accordance with the Documentation, will operate in substantial conformity with the description of the SOFTWARE set forth in the Documentation. The Company does not warrant that the SOFTWARE will meet your requirements or that the operation of the SOFTWARE will be uninterrupted or error-free. The Company warrants that the media on which the SOFTWARE is delivered shall be free from defects in materials and workmanship under normal use for a period of thirty (30) days from the date of your purchase. Your only remedy and the Company's

only obligation under these limited warranties is, at the Company's option, return of the warranted item for a refund of any amounts paid by you or replacement of the item. Any replacement of SOFTWARE or media under the warranties shall not extend the original warranty period. The limited warranty set forth above shall not apply to any SOFTWARE which the Company determines in good faith has been subject to misuse, neglect, improper installation, repair, alteration, or damage by you. EXCEPT FOR THE EXPRESSED WARRANTIES SET FORTH ABOVE, THE COMPANY DISCLAIMS ALL WARRANTIES, EXPRESS OR IMPLIED, INCLUDING WITHOUT LIMITATION, THE IMPLIED WARRANTIES OF MERCHANTABILITY AND FITNESS FOR A PARTICULAR PURPOSE. EXCEPT FOR THE EXPRESS WARRANTY SET FORTH ABOVE, THE COMPANY DOES NOT WARRANT, GUARANTEE, OR MAKE ANY REPRESENTATION REGARDING THE USE OR THE RESULTS OF THE USE OF THE SOFTWARE IN TERMS OF ITS CORRECTNESS, ACCURACY, RELIABILITY, CURRENTNESS, OR OTHERWISE.

IN NO EVENT, SHALL THE COMPANY OR ITS EMPLOYEES, AGENTS, SUPPLIERS, OR CONTRACTORS BE LIABLE FOR ANY INCIDENTAL, INDIRECT, SPECIAL, OR CONSEQUENTIAL DAMAGES ARISING OUT OF OR IN CONNECTION WITH THE LICENSE GRANTED UNDER THIS AGREEMENT, OR FOR LOSS OF USE, LOSS OF DATA, LOSS OF INCOME OR PROFIT, OR OTHER LOSSES, SUSTAINED AS A RESULT OF INJURY TO ANY PERSON, OR LOSS OF OR DAMAGE TO PROPERTY, OR CLAIMS OF THIRD PARTIES, EVEN IF THE COMPANY OR AN AUTHORIZED REPRESENTATIVE OF THE COMPANY HAS BEEN ADVISED OF THE POSSIBILITY OF SUCH DAMAGES. IN NO EVENT SHALL LIABILITY OF THE COMPANY FOR DAMAGES WITH RESPECT TO THE SOFTWARE EXCEED THE AMOUNTS ACTUALLY PAID BY YOU, IF ANY, FOR THE SOFTWARE.

SOME JURISDICTIONS DO NOT ALLOW THE LIMITATION OF IMPLIED WARRANTIES OR LIABILITY FOR INCIDENTAL, INDIRECT, SPECIAL, OR CONSEQUENTIAL DAMAGES, SO THE ABOVE LIMITATIONS MAY NOT ALWAYS APPLY. THE WARRANTIES IN THIS AGREEMENT GIVE YOU SPECIFIC LEGAL RIGHTS AND YOU MAY ALSO HAVE OTHER RIGHTS WHICH VARY IN ACCORDANCE WITH LOCAL LAW.

ACKNOWLEDGMENT

YOU ACKNOWLEDGE THAT YOU HAVE READ THIS AGREEMENT, UNDERSTAND IT, AND AGREE TO BE BOUND BY ITS TERMS AND CONDITIONS. YOU ALSO AGREE THAT THIS AGREEMENT IS THE COMPLETE AND EXCLUSIVE STATEMENT OF THE AGREEMENT BETWEEN YOU AND THE COMPANY AND SUPERSEDES ALL PROPOSALS OR PRIOR AGREEMENTS, ORAL, OR WRITTEN, AND ANY OTHER COMMUNICATIONS BETWEEN YOU AND THE COMPANY OR ANY REPRESENTATIVE OF THE COMPANY RELATING TO THE SUBJECT MATTER OF THIS AGREEMENT.

Should you have any questions concerning this Agreement or if you wish to contact the Company for any reason, please contact in writing at the address below.

Robin Short
Prentice Hall PTR
One Lake Street
Upper Saddle River, New Jersey 07458